Synthesis Lectures on Engineering, Science, and Technology

The focus of this series is general topics, and applications about, and for, engineers and scientists on a wide array of applications, methods and advances. Most titles cover subjects such as professional development, education, and study skills, as well as basic introductory undergraduate material and other topics appropriate for a broader and less technical audience.

Hamid Radmard Rahmani

Seismic Control of Structures Using Tuned Mass Dampers

Theory and Practice

Springer

Hamid Radmard Rahmani
Griesemann Engineering GmbH & Co. KG
Leipzig, Germany

ISSN 2690-0300 ISSN 2690-0327 (electronic)
Synthesis Lectures on Engineering, Science, and Technology
ISBN 978-3-031-71219-7 ISBN 978-3-031-71217-3 (eBook)
https://doi.org/10.1007/978-3-031-71217-3

This Springer imprint is published by the registered company Springer Nature Switzerland AG
The registered company address is: Gewerbestrasse 11, 6330 Cham, Switzerland

To my daughter, Mantreh,
who brings light and inspiration into my life,
and in honor of my father, Ahmed,
a brilliant structural engineer who taught me that
for an engineer, nothing is impossible—only
challenges waiting for solutions.

Contents

List of Figures

List of Tables

List of Algorithms

List of Symbols

A	Acceleration response vector of structure and TMDs
$\ddot{U}$	Acceleration vector
a_i^C	Controlled acceleration response of story i
u_i^C	Controlled displacement response of story i
C	Damping matrix
C_{st}	Damping matrix of the structure
C_t	Damping matrix of the TMD
c_{st}	Damping of the structure
c_t	Damping of the TMD
ζ_n	Damping ratio for nth mode
$\mathbf{u}(t)$	Displacement response of the building
U	Displacement response vector of structure and TMDs
ν_{st}	Frequency of the structure
ν_t	Frequency of the TMD
β	Frequency ratio of the TMD to the structure
$\ddot{u}_g$	Ground acceleration
P_t	Inertial forces
M	Mass matrix
M_{st}	Mass matrix of the structure
M_t	Mass matrix of the TMD
m_{st}	Mass of the structure
m_t	Mass of the TMD
m_0	Mass ratio of the TMD to the structure
$q_n(t)$	Modal coordinate for nth mode
ϕ_{jn}	Modal displacement for mass n in jth mode
m_j	Modal mass for the jth mode
Γ_n	Modal participation ratio
$\mathbf{u}_n(t)$	Mode nth displacement response

ω_n	Natural frequency (undamped) (rad/sec)
N	Number of stories
ϕ_n	nth natural vibration mode
J_i	Objective function i
K	Stiffness matrix
K_{st}	Stiffness matrix of the structure
K_t	Stiffness matrix of the TMD
i	Story number
a_i^{UC}	Uncontrolled acceleration response of story i
u_i^{UC}	Uncontrolled displacement response of story i
v_i^{UC}	Uncontrolled velocity response of story i
V	Velocity response vector of structure and TMDs
$\dot{U}$	Velocity vector

Structural Control

1

1.1 Introduction

Vibration control in structures holds paramount importance in ensuring their safety and performance. Addressing this critical aspect, Tuned Mass Dampers (TMDs) emerge as a vital solution that has been extensively researched and implemented in both practical applications and rigorous testing under real wind and earthquake events. The proven reliability of TMDs further underscores their significance in mitigating vibrations and enhancing the overall structural response.

Amidst the continuous advancements in structural control systems, TMDs consistently stand out as a favored alternative for controlling vibrations. These systems offer a range of advantages, including their ease of installation and cost-effectiveness, making them a versatile and practical choice for countering the effects of wind and earthquake excitations. Their successful implementation and rigorous testing in real-world scenarios validate their efficacy and enhance their credibility as a reliable vibration control solution.

The author's experience in research and practice sides of structural engineering indicates that although many excellent books exist on the topic, it has often been challenging to find an optimal way to learn the underlying theories. Most state-of-the-art books assume readers are already familiar with the relevant concepts, and although basic books cover the theories, it is time-consuming to sift through them all to determine which ones are applicable to specific research projects. Additionally, the practical implementation of these theories is often overlooked. Dealing with these issues can be time-consuming and adversely impact the efficiency of research efforts. Therefore, gaining insight into recent advances and quickly understanding their implementation would greatly benefit researchers and engineers, allowing them to make informed decisions about delving deeper into specific areas.

Recognizing the need for a comprehensive resource that bridges the gap between theoretical knowledge and practical applications, this book endeavors to provide a holistic understanding of TMDs. It caters to the requirements of structural engineers, researchers, and

H. Radmard Rahmani, *Seismic Control of Structures Using Tuned Mass Dampers*,
Synthesis Lectures on Engineering, Science, and Technology,
https://doi.org/10.1007/978-3-031-71217-3_1

students pursuing advanced degrees, empowering them to leverage TMDs in their projects and contribute to further advancements in this field.

The book initiates by providing a concise yet thorough overview of structural control techniques, elucidating their underlying principles and highlighting their practical utility. It also delves into the key factors that should be considered when determining the suitability of such systems for a given structure.

Building upon this foundation, the subsequent chapter explores the latest advances in TMD technology, shedding light on the current state of research and development. The chapter not only offers valuable insights into cutting-edge methodologies but also presents real-world case studies showcasing the successful application of TMDs. This holistic perspective equips researchers with inspiration for potential developments and enables structural engineers to readily adopt state-of-the-art techniques in their projects.

In addition, the book dedicates a comprehensive chapter to various structural analysis methods, providing a framework for engineers to make informed decisions based on the specific requirements of their projects. Furthermore, it introduces a MATLAB code specifically developed for dynamic analysis, which is subsequently utilized in the final chapter to evaluate the performance of high-rise buildings subjected to earthquake excitations. The chapter also explores the implementation of static condensation, an effective technique for reducing the size of structural matrices, thereby optimizing analysis efficiency.

Central to the book's objective is chapter five, which meticulously covers the theories, modeling techniques, and analysis methodologies specific to TMDs. It presents readers with a deep understanding of how TMDs are accurately modeled and rigorously analyzed, while also providing practical considerations for designing optimal TMD parameters. Crucially, this chapter outlines the main criteria for evaluating the performance of TMD systems, differentiating between wind and earthquake excitations.

Finally, the book concludes with a comprehensive case study in chapter six, guiding readers through the step-by-step process of designing a TMD to enhance the seismic response of a high-rise building. By illustrating the complete workflow from structural modeling and analysis to decision-making on TMD parameters and performance evaluation, this chapter serves as a practical guide for readers seeking to implement TMDs in real-world scenarios.

Throughout the book, the integration of MATLAB code for structural analysis enables readers to validate their findings and compare results with established commercial software such as CSI SAP2000. This hands-on approach enhances the practicality of the book, enabling readers to effectively apply analysis modules and gain valuable insights into their real-world application.

1.2 Who Can Benefit from This Book?

This book serves as a valuable practical guide tailored for PhD and master's students seeking to integrate Tuned Mass Dampers (TMDs) into their research endeavors. Additionally, it proves beneficial for engineers aiming to incorporate TMDs into real-world projects or expand their understanding of the advantages and limitations associated with these systems. While the author's primary objective is to offer practical content, it is important to note that this book also provides a concise overview of the current state of research and relevant theories. Consequently, readers seeking more comprehensive theoretical materials on TMDs may choose to explore the referenced sources for further in-depth exploration.

1.3 How to Read the Book

This book has been structured to provide readers with flexibility in navigating and extracting relevant information based on their specific needs. Here is a breakdown of how readers can effectively engage with the different demand:

Researchers/Students: Researchers and students pursuing PhD or master's degrees can benefit from gaining in-depth knowledge and exploring the theoretical aspects of TMDs. The following chapters are particularly relevant to this group:

- Overview and Importance of Structural Control Techniques (In this chapter)
- Latest Advances in Tuned Mass Dampers (Chap. 2)
- Applicable Theories (Chap. 3)
- MATLAB Code and Analysis Modules (Appendix)

These chapters provide a solid foundation in understanding the theoretical principles, advancements, and evaluation methods related to TMDs. They offer valuable insights into the research landscape, enabling researchers/students to contribute to the field and advance their understanding of TMD technology.

Engineers: Engineers involved in real-world projects or seeking to apply TMDs in practical scenarios can benefit from chapters that provide practical guidance, design examples, and analysis methodologies. The following chapters are particularly relevant to engineers:

- Modeling, Analysis, and Performance Evaluation of TMDs (Chap. 4)
- Practical Design Example: TMDs for Seismic Response Improvement (Chap. 5)

These chapters offer practical insights, decision-making criteria, and step-by-step guidance for engineers looking to utilize TMDs in their projects. They provide essential information on structural analysis methods, design considerations, performance evaluation, and practical implementation using MATLAB code. It's important to note that while the categorization above indicates the chapters more relevant to each group, all chapters contain valuable information that can benefit both researchers/students and engineers. The book's modular structure allows readers to select chapters based on their specific interests and requirements, ensuring a tailored reading experience.

1.4 An Overview of Structural Control Systems

1.4.1 Introduction

Vibration control poses a significant challenge in structural engineering, as it involves dealing with various uncertainties. Environmental loads such as earthquakes and winds exhibit stochastic behavior, while the inherent complexity of material behavior adds further challenges. Moreover, deterministic analysis methods often yield results that differ from real-world observations, contributing to uncertainties in predicting structural behavior.

Traditional approaches rely solely on the inherent capacity of structures to withstand environmental loads. While effective for many structures, advancements in engineering have led to the development of increasingly complex structures, which demand higher levels of performance and safety. This has created a pressing need for alternative methods that go beyond classical approaches, particularly in the construction of high-rise buildings, long-span bridges, and specialized facilities like nuclear installations.

Structural control systems have emerged as a modern solution to address these challenges, with the concept initially introduced by Yao in 1972 [9]. Since then, extensive research has been conducted on these systems, resulting in the development of various types of structural control methods. The following sections aim to provide an overview of these systems, elucidating their purpose, significance, and the circumstances that necessitate their implementation.

Structural control systems offer a proactive approach to vibration control, allowing engineers to actively manage and mitigate the effects of dynamic loads on structures. By employing these systems, engineers can enhance structural performance, increase safety margins, and optimize cost-effectiveness in the face of challenging environmental conditions.

The subsequent sections will delve into the different types of structural control systems, examining their underlying principles, implementation considerations, and benefits. The goal is to provide readers with a comprehensive understanding of these systems, enabling them to make informed decisions about when and how to utilize structural control methods in their projects. By exploring the range of available options, readers will gain valuable insights into selecting the most suitable control systems for addressing specific vibration control challenges in structural engineering.

1.4.2 Seismic Behavior of Structures

In the context of structural engineering, the term **behavior** refers to the dynamic response of a building or structure when subjected to seismic (earthquake-induced) excitations. The seismic behavior of a structure encompasses a wide range of complex interactions between its components, materials, and the ground motion. Understanding and analyzing the seismic behavior of structures is essential for ensuring their safety, performance, and resilience in earthquake-prone regions.

When an earthquake occurs, the ground motion imparts various forces and accelerations to a structure. These forces trigger deformations, vibrations, and potentially damaging effects within the building. The seismic behavior of a structure is characterized by its ability to withstand these forces and continue to function as intended without experiencing catastrophic failure.

Seismic behavior encompasses several critical aspects:

1. Resonance and Natural Frequencies: Structures have inherent natural frequencies at which they tend to vibrate when excited. If the frequency of the ground motion matches the natural frequency of the structure, resonance can occur, leading to amplified vibrations and potential structural damage. Understanding the natural frequencies and modes of vibration is crucial for avoiding resonance effects.
2. Deformation and Displacement: Seismic forces induce deformations and displacements in a structure. Excessive displacements can result in structural damage, affecting the building's stability and functionality. Analyzing the expected displacements and ensuring they remain within acceptable limits is a key consideration in seismic design.
3. Stress and Strain: Seismic forces subject structural components to stress and strain. Excessive stress can lead to material failure, while large strains can impair the structural integrity. Ensuring that stress and strain levels remain within safe limits is vital to prevent structural collapse.
4. Damping and Energy Dissipation: Damping refers to the ability of a structure to absorb and dissipate energy during vibration. Proper damping mechanisms, such as tuned mass dampers (TMDs), can reduce the effects of resonance and prevent excessive vibrations.
5. Interaction with Foundation and Soil: The behavior of a structure is influenced by the interaction between the building's foundation and the surrounding soil. Soft soils can amplify ground motion, while liquefaction can significantly compromise the stability of a structure's foundation.
6. Dynamic Response Spectrum: The dynamic response spectrum provides a graphical representation of the expected structural response to varying frequencies of ground motion. Analyzing the response spectrum helps engineers assess the potential vulnerability of a structure to different types of earthquakes.

By comprehensively studying and analyzing these aspects of seismic behavior, engineers can develop effective strategies for designing and retrofitting structures to withstand earthquake forces. This includes optimizing structural components, incorporating damping systems, and ensuring that the building can safely absorb and dissipate energy during seismic events. Ultimately, a thorough understanding of seismic behavior is essential for creating resilient and earthquake-resistant structures that can protect lives and property in seismic-prone regions.

1.4.3 Structural Control Systems

The utilization of a structural control system aims to enhance the structural responses to external excitations such as wind or earthquakes. This improvement is guided by three key aspects outlined in building codes for assessments and verification: stability, strength, and serviceability.

Stability: The stability criteria encompass both global and local levels. Global instabilities, such as buckling, sliding, and overturning, must be addressed to ensure the overall stability of the structure. Similarly, local instabilities, such as flexural-torsional buckling in structural elements, should be considered to prevent localized failures.

Strength: The strength criteria focus on ensuring that the structural members can withstand the resultant stresses under critical load combinations. This involves comparing the resultant stresses and forces under such load combinations with the corresponding allowable values to verify that the strength requirements are met.

Serviceability: Serviceability criteria ensure that the structure remains functional or partially functional during critical loading situations. In a performance-based design approach, serviceability is a crucial aspect that governs the performance level of the structure. For instance, structures with specific functions like hospitals may require an operational (OP) performance level, indicating that they should remain fully functional following external excitations such as earthquakes. Another example involves the lateral movements of tall buildings under wind loads. While the structure may remain stable under these loads, excessive movements beyond certain limits may lead to occupant discomfort and a loss of functionality.

In certain regions, structural engineers face the challenge of designing structures subjected to high dynamic loads, such as cities located near fault lines prone to strong earthquakes or offshore areas experiencing intense winds. In such locations, traditional structural design approaches may prove to be impractical, challenging, or costly in meeting one or more of the three essential structural design criteria. In these cases, employing a structural control system can offer a viable solution to satisfy the design criteria.

The utilization of a structural control system becomes a logical choice in various circumstances, which are further explored in the subsequent section. These circumstances arise when conventional design methods alone are insufficient or uneconomic to fulfill the required stability, strength, and serviceability criteria. Generally speaking, the development of structural control systems aims to either reduce the demand on the primary structure or increase its capacity to withstand dynamic loads. The terms "Demand" and "Capacity" play crucial roles in understanding the behavior of structures and are integral to performance-based design approaches. By delving deeper into these concepts, engineers gain greater control over the design criteria and can explore the intricacies of structural behavior.

In essence, the **demand** pertains to the deformations and loads that the structure needs to withstand, while the **capacity** represents the structure's ability to resist such loads and deformations, considering all three aforementioned design criteria. Understanding the interplay between demand and capacity is essential for ensuring the structural integrity, stability, and functionality of the designed system. Through comprehensive assessments of both demand and capacity, engineers can make informed decisions to optimize the performance and reliability of structures under dynamic loading conditions. For instance, base isolation systems are designed to reduce the demand by subjecting the structure to lower displacements during an earthquake. On the other hand, hydraulic dampers are employed to increase the capacity of the structure by dissipating energy. These examples highlight different strategies employed in structural control systems to effectively manage the response of structures to dynamic loads.

Depending on the need for external power supply, the passive structural controllers are classified into **passive**, **active**, **semi-active**, and **hybrid** systems. Passive control systems were the first to be implemented historically and offer the advantage of not requiring any external power supply. Extensive research has been conducted on passive controllers [10–14] and they have been successfully employed in numerous structures worldwide. Examples of passive control techniques include base isolators, tuned mass dampers, liquid dampers, and friction dampers. Some common types of passive dampers are illustrated in Fig. 1.1. Passive systems are frequently the initial consideration for structural engineers when exploring different types of controllers. This preference stems from their advantages in terms of design simplicity, ease of implementation, and lower maintenance costs, making them an appealing option in structural control systems.

As illustrated in Fig. 1.2, passive control systems, such as dampers, exert control forces on the structure that are predominantly influenced by the structural responses rather than external excitations. This establishes an indirect relationship between the control forces and the external excitations.

$$F = f(R) \tag{1.1}$$

Where F refers to the central forces and R refers to structural responses

However, passive systems have a drawback in that their parameters are constant, limiting their adaptability to the external excitations compared to other types of control systems. For

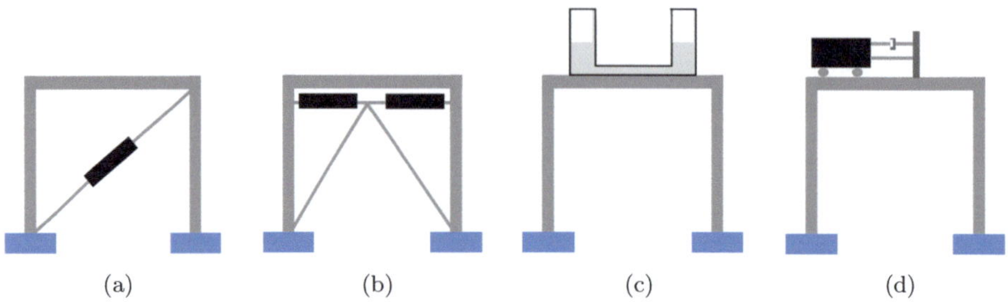

Fig. 1.1 Common types of dampers: **a** Diagonal brace and viscous damper, **b** Chevron brace and viscous damper, **c** Tuned liquid column damper, **d** Tuned mass damper

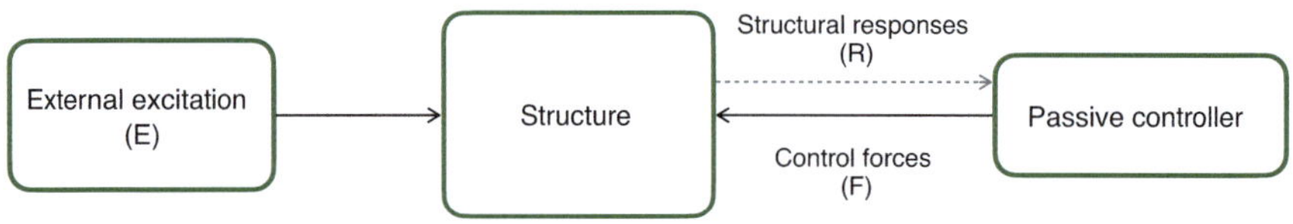

Fig. 1.2 Passive control systems

instance, they cannot differentiate between wind and earthquake excitations and adjust their behavior accordingly.

Active controllers, on the other hand, overcome this limitation by incorporating sensors to monitor external excitations in real-time. By actively sensing and analyzing the structural responses, these controllers can dynamically adjust the control forces to optimize the structural performance [15–19]. As it is shown in the Fig. 1.3, in active control, the external excitations can directly participate in determining the control forces allowing more adaptability. These systems can be classified as **feedback** or **feedforward** control, depending on whether the external excitations are considered in the determination of control forces.

In a feedback control system, the structural responses are continuously monitored and fed back into the control function. This allows for real-time adjustments based on the actual behavior of the structure, leading to a more adaptive and responsive control strategy. By incorporating the structural responses, the control system can actively counteract the effects of external excitations and mitigate their impact on the structure.

$$F = f(E, R) \tag{1.2}$$

Where E refers to external excitations

On the other hand, feedforward control systems primarily rely on predictive models and anticipated external excitations to determine the control forces. These models take into account the expected behavior of the structure under specific excitations, allowing for proactive control actions to be taken in advance. By anticipating the external excitations,

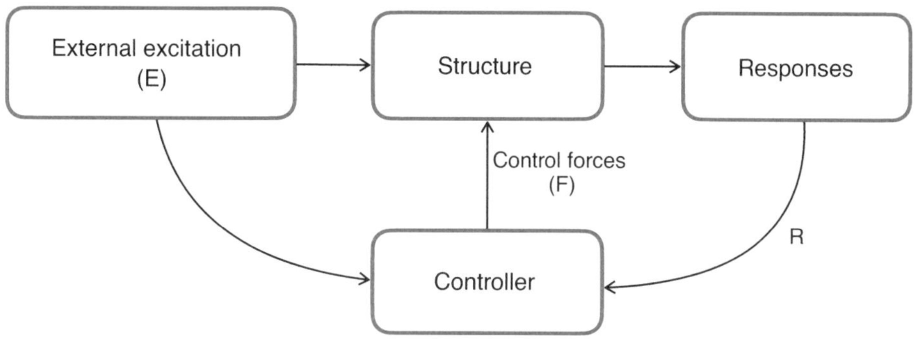

Fig. 1.3 Active control systems

feedforward control systems can effectively minimize the structural response and enhance the overall performance.

However, one of the disadvantages of active control systems is their reliance on a continuous power supply to operate effectively. This requirement for power can pose challenges, particularly during natural hazards such as earthquakes where power infrastructure may be compromised. Interruptions in the power supply can limit the functionality of active control systems and hinder their ability to actively respond to dynamic loads.

To address this limitation, semi-active control systems offer an alternative solution. These systems require a lower power supply compared to active control systems and can operate using batteries or power storage systems. By modifying the parameters of passive controllers in real-time, such as adjusting the stiffness of a tuned mass damper, semi-active control systems can optimize their performance in mitigating structural vibrations and improving the overall response of the structure. The reduced power requirement of semi-active control systems enhances their feasibility and applicability in situations where a continuous power supply may not be readily available [20–23].

Hybrid systems combine different types of controllers to create an integrated and versatile control system. The purpose of designing a hybrid system is incorporating both active and passive control elements to enhance the efficiency and adaptability of the overall control system. One of the key advantages of hybrid systems is their ability to switch between active and passive modes, depending on the specific requirements of the structure [24].

It is important to acknowledge that as we move from passive systems towards hybrid systems, the design of the control system becomes increasingly complex and challenging. Structural engineers are confronted with a multitude of complexities and uncertainties in order to successfully implement hybrid systems. The integration of different control strategies, the coordination of active and passive components, and the optimization of their interactions present additional design considerations that require careful analysis and decision-making.

In hybrid systems, the engineer must carefully balance the benefits of combining different control techniques with the associated challenges. This includes considering the compatibility and synchronization of various control elements, addressing potential conflicts or inconsistencies between active and passive components, and ensuring the overall system operates seamlessly and effectively. Furthermore, the uncertainties related to the dynamic behavior of the structure and the varying external loads introduce additional design challenges that need to be addressed.

Understanding the distinctions between the different types of structural control systems enables engineers and researchers to make informed decisions when selecting the most appropriate approach for a given project. Each type, including passive, active, semi-active, and hybrid systems, has its advantages, considerations, and suitability depending on factors such as the desired level of adaptability, power availability, maintenance requirements, and specific project constraints.

1.4.4 When to Use a Structural Control System

Structural control systems are considered in various scenarios where traditional design approaches may not be sufficient or practical. In practice, the evaluation of the need for implementing a control system are done based on several key factors:

Resonance: Resonance occurs when the frequency of external excitation aligns closely with the natural frequency of the structure in a specific mode. This phenomenon can lead to significant displacement responses and potential structural instabilities. By incorporating damping mechanisms, structural control systems can mitigate the effects of resonance, ensuring the safety and performance of the structure. The collapse of the Tacoma Narrows bridge in 1940 (Fig. 2.13), driven by torsional resonance induced by wind excitations, serves as an illustrative example of the importance of controlling resonance effects.

High Natural Frequency: Structures with high natural frequencies in the horizontal directions often exhibit lower lateral stiffness. This is commonly observed in high-rise and tall buildings. Implementing structural control systems in such cases can improve the structural response, enhance stability, and minimize excessive movements during external excitations.

Architectural Provisions: When architectural aspects restricts the provision of adequate lateral load-bearing systems, structural control systems can offer an alternative solution. These systems can effectively mitigate the dynamic response of the structure, ensuring its stability and performance.

Serviceability Criteria: In situations where meeting the serviceability criteria solely based on the inherent capacity of the structure is challenging or economically unfeasible, structural control systems become a viable option. For example, while the structural strength may be sufficient to withstand lateral loads, the drift or deformations may exceed allowable values according to the code. Additionally, there may be limitations on providing the required ductility due to deformation limits in non-structural elements such as architectural components or installations. In such cases, implementing a structural control system can effectively manage and control these movements, ensuring strict adherence to the serviceability requirements and preserving the integrity and functionality of the structure.

Structural Characteristics: In some cases, special structural characteristics, such as inverted pendulum structures, pose challenges in providing conventional lateral load-bearing systems. Structural control systems offer an alternative approach to ensure stability and control for such structures, addressing their specific requirements effectively.

High Importance Level: Certain structures hold critical importance where failure or collapse would result in significant life and financial losses. Examples include tall buildings and essential facilities like hospitals, military installations, and governmental buildings. These structures often require operational functionality even during natural hazards. By incorporating an additional load-bearing system that activates during unexpected lateral excitations, structural engineers can meet the demanding safety, durability, and serviceability criteria for these structures.

By understanding these considerations, structural engineers can assess the suitability of employing a structural control system in specific design scenarios. The subsequent sections will delve into the design, implementation, and performance aspects of these control systems, providing valuable insights and guidance for informed decision-making in real-world applications.

1.4.5 Choosing a Structural Control System

Choosing the appropriate structural control system is a challenging task that involves various factors to be considered in the early stages of the design process. While a comprehensive discussion on this topic would require an entire book, we will highlight some of the most important factors to guide the decision-making process.

In practical terms, passive systems are typically the first type to be considered. This is primarily due to their cost-effectiveness, ease of installation and design, and the fact that they do not require an external power supply. Passive systems also offer the advantage of lower maintenance costs and a track record of successful implementation in numerous real-world projects worldwide. Thus, they present a reliable and economical option that should be given serious consideration during the structural planning phase.

However, in more complex situations, other control systems such as active, semi-active, or hybrid systems may need to be explored. For instance, if the structure is located in an area prone to both high winds and strong earthquakes, a passive control system with constant parameters may not provide an optimal solution for both scenarios. In such cases, an adaptive system that can adjust its behavior based on the specific external excitations may be considered as an alternative. In some complex scenarios, the optimal solution may involve a combination of different passive systems. For example, employing two mass dampers tuned to different frequencies to address the specific external excitations.

Architectural and geometrical constraints also play a crucial role in the selection of a structural control system. The design process requires close collaboration between the structural engineer and the architect to ensure the integration of the control system within the overall architectural plan. For instance, when utilizing mass dampers, careful consideration should be given to the placement of these dampers, often in the top stories and close to the center of mass of each story, as reflected in the architectural plans.

Costs and construction time are additional important factors that influence the decision-making process. Typically, multiple control system options may be applicable to a project, but they may differ in terms of required time, cost, and modifications to the architectural plan. Consequently, the structural engineer evaluates two or three alternatives during the initial design phase, taking into account these factors to determine the most suitable solution.

Although it is not possible to provide a step-by-step procedure for choosing a control system, the following steps are commonly taken by structural engineers to gather useful information and facilitate decision-making regarding the control system:

1. Determining project specifications and limits: This involves gathering information about the building's geometry, function, location, external loads, architectural and geotechnical specifications and limits, as well as the project budget.
2. Evaluating the maximum inherent capacity of the structure: Conducting preliminary analyses to identify the main issues in the absence of a control system provides insight into the structural limitations.
3. Developing an improvement plan: Based on the identified problems, the structural engineer prepares a plan that clearly defines the purpose of implementing a structural control system. It addresses questions such as which aspects need to be improved whether it is reducing deformations, increasing stability, addressing load uncertainties, or modifying the structure's natural frequencies.
4. Quantifying the problem: This step involves delving into more detailed analysis and collaborating closely with architectural and geotechnical teams to quantify the problems. By assigning values to the capacity and demand, a clearer understanding of the issues is obtained. This step requires in-depth analysis to minimize problems as much as possible without resorting to advanced and complicated procedures.
5. Suggesting possible control systems: Considering the type and magnitude of the problem, the structural engineer suggests at least two alternatives and compares them. Experience plays a crucial role in this step, aiming to present feasible and practical structural solutions that take into account economical, structural, and architectural factors for the client. Advanced and detailed analysis is not yet performed at this stage, as the focus is on major parameters.
6. Approval of the control system: In this phase, a structural control system is initially approved. Initial comments and concerns from all involved professionals in the project

are gathered and discussed to reach a feasible initial plan. Typically, some adaptations and updates to the plans are required at this stage.

7. Structural design: The structural engineer designs the controller parameters by conducting structural modeling and analysis, providing an overview of the size and positioning of the main structural elements and the specifications of the control system. This information is then shared with other departments for review, updates, and feedback to ensure the practicality of the final structural system.
8. Final adjustments and detailing: In this final phase, the control system undergoes final adjustments, and detailed design work is carried out. Additionally, execution guides and construction plans are prepared to facilitate the implementation of the control system.

By following these steps, structural engineers can gather relevant information, analyze the project's requirements, and make informed decisions to select an appropriate structural control system that meets the project's objectives and constraints.

Tuned Mass Dampers

2

2.1 Introduction

A Tuned Mass Damper (TMD) is a mechanical energy dissipating system which in its simplest form consists of a mass which is attached to the structure by a spring, and a dashpot (see Fig. 2.1). The utilization of TMDs has emerged as a prominent solution for controlling vibrations in buildings and structures. Vibrations induced by various factors such as wind, earthquakes, and dynamic loads can lead to discomfort for occupants, structural fatigue, and even compromise the safety and functionality of the building. TMDs offer an effective means of mitigating these vibrations by introducing a secondary mass and associated damping system that counteracts the building's oscillations.

TMDs operate on the principle of dynamic vibration absorption. By incorporating an additional mass connected to the primary structure through springs and dampers, TMDs are designed to resonate at specific frequencies corresponding to the building's natural modes of vibration. This resonance allows the TMD to actively counteract and suppress the unwanted vibrations, reducing their amplitudes and enhancing the overall structural stability.

The first concept of TMDs was applied by Frahm in 1909 [25] to mitigate the rolling motion of a ship. Since that time, TMDs were widely utilized in the vibration control of dynamic systems. In structural engineering, the main utilization area of the TMDs belongs to the bridges and buildings. In recent years, TMDs have gained significant attention in the field of structural engineering due to their proven effectiveness in vibration control. They have been successfully employed in a wide range of building types, including high-rise structures, bridges, stadiums, and even historical landmarks. TMDs have demonstrated their ability to mitigate vibrations caused by a variety of factors, including wind-induced oscillations, seismic events, and human-induced vibrations.

The key advantage of TMDs lies in their ability to provide targeted vibration control. Through careful design and tuning, TMDs can effectively address specific vibration modes

H. Radmard Rahmani, *Seismic Control of Structures Using Tuned Mass Dampers*,
Synthesis Lectures on Engineering, Science, and Technology,
https://doi.org/10.1007/978-3-031-71217-3_2

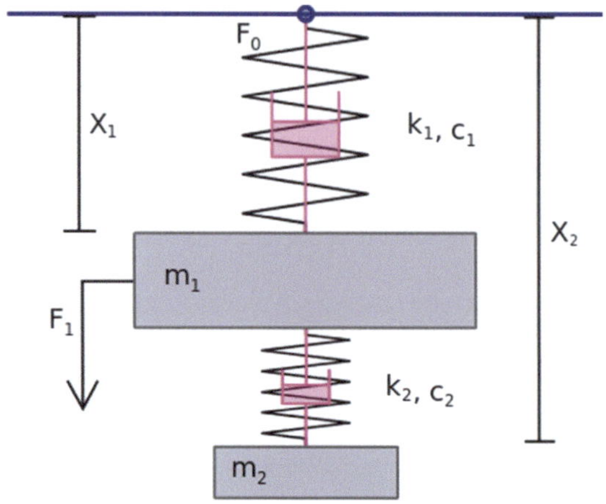

Fig. 2.1 Tuned mass damper attached to a single degree of freedom system. m_1, k_1, c_1: Structure's mass stifness and damping. m_2, k_2, c_2: TMD's mass, stifness, and damping

that are of concern, offering a tailored solution for each building's unique dynamics. This adaptability makes TMDs particularly valuable in situations where conventional passive damping methods may not be sufficient or practical.

Moreover, TMDs offer several benefits over alternative vibration control strategies. They can be used to retrofit the existing structures, making them a viable option for improving the performance of older buildings. TMDs are also cost-effective compared to other active control systems, as they do not require a continuous external power supply or complex control algorithms. By understanding the principles and design considerations of TMDs, structural engineers can effectively implement these devices to mitigate vibrations and optimize the behavior of buildings in various environmental conditions.

The utilization of TMDs in various types of bridges, including pedestrian, railway, and transporter bridges, has attracted significant attention in research and engineering [26–30]. TMDs play a crucial role in mitigating vibrations caused by dynamic loads such as pedestrian-induced forces and wind loads.

Pedestrian bridges often experience resonance due to the dynamic loads imposed by human movement or gusty winds. TMDs have proven to be an effective solution in addressing these resonance issues. By introducing an additional mass connected to the bridge structure through springs and dampers, TMDs can actively counteract the vibrations induced by pedestrians or wind, reducing their impact and improving the overall stability and comfort of the bridge.

Another critical application of TMDs is in controlling deck movements caused by wind loads in suspension bridges. The dynamic loads generated by wind can coincide with one of the bridge's natural modes, leading to excessive displacements, instabilities, and potential failures of bridge members. In such cases, TMDs are strategically tuned to the critical natural frequency of the bridge, enabling them to effectively control the corresponding modal responses and mitigate the adverse effects of wind-induced vibrations.

Figure 2.2 illustrates that TMDs can be implemented for vertical or horizontal acting configurations, depending on the critical mode shape of the bridge. The optimal placement of the TMD is determined by structural engineers based on comprehensive analysis results, aiming to position it at the location with the maximum modal displacement (as shown in Fig. 2.3). The utilization of TMDs in bridge design and retrofitting projects demonstrates their effectiveness in improving the overall structural behavior and ensuring the long-term durability of bridges subjected to dynamic loads.

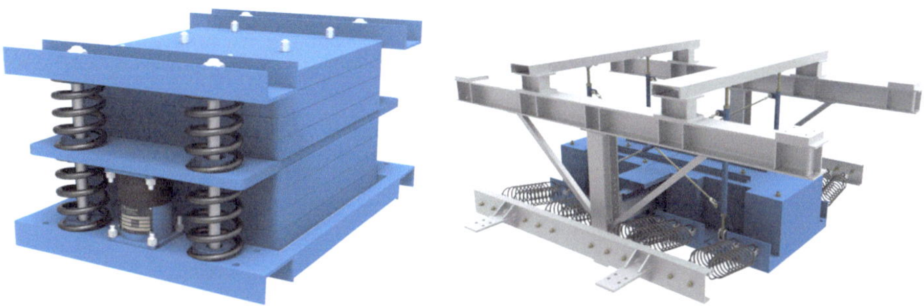

(a) (b)

Fig. 2.2 Tuned mass dampers for vibration control in bridges: **a** Vertically acting TMD, **b** Horizontal acting TMD [1]

(a) (b)

Fig. 2.3 Application of TMDs in the bridges [1]

Fig. 2.4 Matagarup pedestrian bridge in Perth, Australia. A TMD with about 15 ton is utilized to control the first lateral mode

The Matagarup Bridge in Perth, Australia is an example of the application of TMDs for vibration control in pedestrian bridges. Situated in the southwest of Australia, east of East Perth, and spanning the Swan River, this impressive bridge measures 370 meters in length and 9 meters in width. It is an architectural marvel, featuring a sophisticated design comprising three large arches.To address pedestrian-induced vibrations, a tuned mass damper were designed which was capable of adjustment within the critical frequency range of 0.4–0.7 Hz. A single TMD, weighing approximately 15 tons, was utilized to mitigate vibrations associated with the first lateral natural frequency. By incorporating an eddy current damper, the TMD's characteristics, including its frequency and damping properties, can be maintained regardless of temperature fluctuations. This eliminates the need to account for changes in absorber frequency and damping level typically associated with viscoelastic dampers during the design process [1] (Fig. 2.4).

In this context, the utilization of TMDs in vibration control of buildings represents a valuable approach to ensure occupant comfort, protect structural integrity, and enhance the overall performance of structures subjected to dynamic loads. Tuned mass dampers have widely implemented for vibration control of buildings such as skyscrapers and high-rise buildings, airport towers, chimneys, and television towers in both research and practice [31–39]. The previous real experiences show that TMDs can effectively reduce the peak responses when the external excitation has large magnitude in one of the structure's natural frequency and produces resonance. For vibration control of buildings, several types of tuned mass dampers (TMDs) are commonly utilized. The specific type of TMD chosen depends on the building's characteristics, vibration modes, and design requirements. Here are some commonly used TMD configurations:

1. **Tuned Mass Pendulum Damper:** This type of TMD consists of a pendulum attached to the structure's floor or roof. The pendulum's mass, length, and damping are tuned to the desired vibration frequency, allowing it to counteract the building's oscillations and reduce vibrations (Fig. 2.5).

Taipei 101 is one of the famous example for utilization pendulum TMDs in buildings. Taipei 101 is located in Taipei, Taiwan, and is renowned for its iconic architecture and impressive height. Standing at 508 meters tall, it held the title of the world's tallest building upon its completion in 2004. To ensure its structural stability and mitigate the effects of wind-induced vibrations, Taipei 101 incorporates a highly sophisticated TMD system (Fig. 2.6).

The TMD used in Taipei 101 is one of the largest and heaviest of its kind in the world. Known as the "Damper Baby," it consists of a massive spherical steel pendulum suspended from the 92nd floor to the 88th floor of the building. This pendulum weighs around 660 metric tons (728 US tons) and is designed to counteract the building's swaying motion caused by wind forces.

The primary purpose of the TMD is to reduce the building's lateral displacements and minimize discomfort for occupants during high winds or seismic events. By employing the principles of tuned mass damping, the Damper Baby operates at a specific frequency that effectively dampens and absorbs the building's vibrations. It functions as a passive control system, relying on the natural frequency of the pendulum to counterbalance the lateral forces acting on the building.

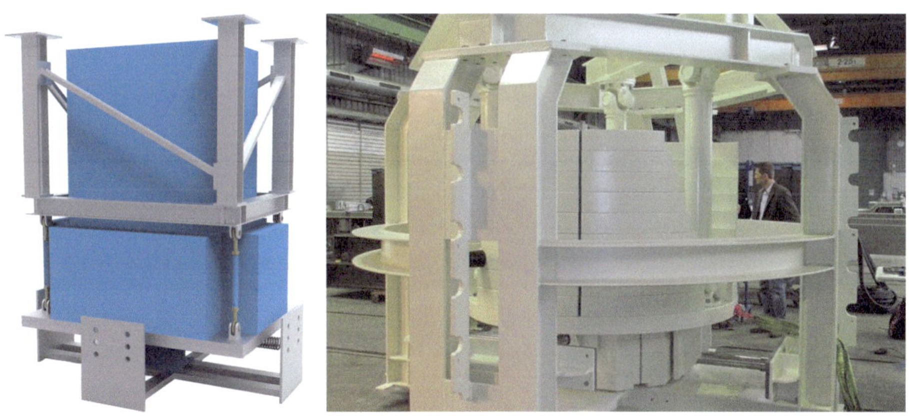

(a) (b)

Fig. 2.5 TMDs with pendulum masses for vibration control of buildings [1]

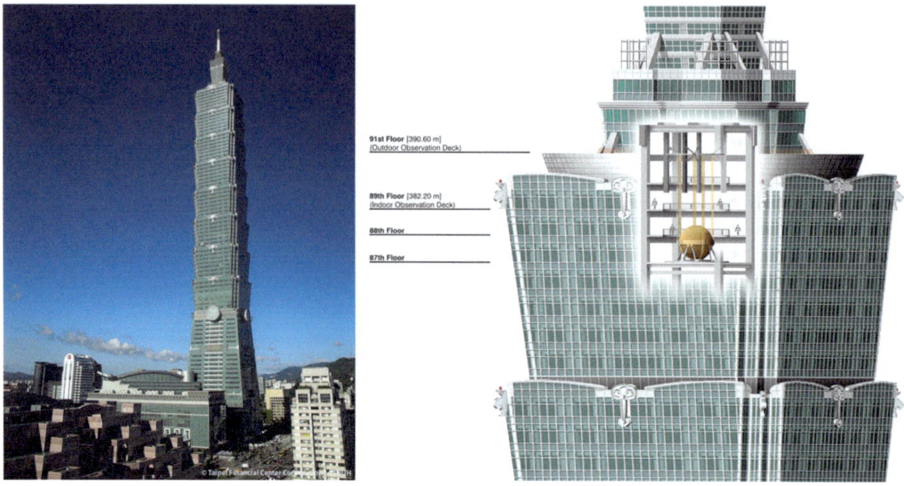

Fig. 2.6 Taipei 101's tuned mass damper

The Damper Baby's movements are carefully coordinated with the building's sway, ensuring that the TMD reacts appropriately to the wind forces. Its position and mass distribution are precisely calibrated to achieve optimal performance. As a result, Taipei 101 experiences significantly reduced lateral displacements, enhancing the overall safety, comfort, and stability of the structure.

The TMD system in Taipei 101 serves as a remarkable example of innovative engineering and design. It has become an iconic feature of the building, attracting visitors who are fascinated by its immense scale and functionality. The Damper Baby showcases the effectiveness of advanced structural control systems in addressing the challenges posed by tall buildings in high-wind areas, further solidifying Taipei 101's status as a remarkable architectural achievement.

2. **Tuned Mass Spring Damper:** In this configuration, a mass is connected to the building via a spring and damper system. The parameters of the spring and damper, such as stiffness and damping, are adjusted to match the building's natural frequency. The TMD counteracts vibrations by absorbing and dissipating energy through the spring and damper mechanism.

One notable example of utilizing Tuned Mass Spring Dampers (TMSD) in buildings is the Citicorp Center (now known as 601 Lexington Avenue) in New York City, United States. This skyscraper, completed in 1977, is a prominent landmark and a prime example of incorporating innovative engineering solutions to address structural challenges. The Citicorp Center features a unique structural design that includes TMSD to mitigate wind-induced vibrations and enhance its overall stability. The building's distinctive feature is its diagonally sloping roof, which creates a visually striking architectural profile. However, this design presented a significant engineering challenge as it made the building susceptible to powerful wind forces.

To address this challenge, the engineers implemented a TMSD system within the building's structural framework. The TMSD in Citicorp Center consists of massive concrete blocks placed on the building's 59th floor. Each block is supported by steel coil springs, which provide the necessary flexibility and damping properties. The TMSD system is specifically tuned to the natural frequency of the building, allowing it to counteract the wind-induced vibrations. As wind forces push and sway the building, the TMSD absorbs and dissipates the energy, effectively reducing the building's lateral movements and enhancing its overall stability.

What makes the TMSD system in Citicorp Center particularly remarkable is the ingenious engineering decision made during its construction. A design flaw in the original plans was discovered after construction had begun, revealing a vulnerability that could potentially cause catastrophic failure in the event of a severe storm. The building's engineers, led by William LeMessurier, implemented an urgent and unprecedented retrofitting process to reinforce the building and strengthen its structural integrity.

The TMSD system played a critical role in this retrofitting process. The concrete blocks on the 59th floor were retrofitted with additional support mechanisms to further enhance the building's ability to withstand extreme wind events. This proactive approach to addressing structural vulnerabilities demonstrated the effectiveness and adaptability of the TMSD system. The implementation of Tuned Mass Spring Dampers in the Citicorp Center showcases the ingenuity and engineering prowess required to address complex structural challenges in high-rise buildings. It stands as a testament to the effectiveness of these passive control systems in enhancing the stability, safety, and performance of tall structures in dynamic environments (Fig. 2.7).

3. **Multiple Tuned Mass Dampers:** In some cases, multiple TMDs are strategically placed at different locations within the building to effectively control vibrations across multiple modes. Each TMD is individually tuned to target a specific vibration mode and reduce its amplitude.
4. **Hybrid TMD Systems:** Hybrid TMD systems combine different types of dampers, such as a combination of tuned mass pendulums and tuned liquid column dampers, to provide enhanced vibration control over a wider frequency range. These systems leverage the unique advantages of each damper type to optimize performance.

The selection of the appropriate TMD configuration depends on factors such as the building's dynamic characteristics, desired vibration reduction targets, available space for installation, and budget constraints. Structural engineers analyze the building's response characteristics and perform simulations to determine the most effective TMD system for vibration control in a given building scenario.

(a) (b)

Fig. 2.7 Citicorp Center in New York equipped with a TMSD to mitigate the wind induced vibrations

2.2 Elements of Tuned Mass Dampers

2.2.1 Mass

Tuned mass dampers (TMDs) typically employ additional masses specifically designed and adjusted to counteract vibrations in a primary structure. These masses are commonly referred to as "secondary masses" or "tuning masses."

The secondary masses utilized in TMDs can vary in size, shape, and material composition, depending on the particular application and engineering specifications. They can range from simple steel or concrete blocks to more intricate designs involving multiple masses, pendulums, or arrays of pendulums. The positioning of the masses along the structure may differ, such as at the top, bottom, or mid-height, depending on the targeted vibration mode.

In essence, the mass within a TMD regulates the maximum magnitude of the control force. When subjected to external excitations, the mass multiplied by the induced acceleration applies control forces to the structure. It should be noted that a heavier mass in a TMD results in larger control forces, which impacts the load-bearing capacity and distribution of the structural elements, transmitting it to the ground. Consequently, the mass size of a TMD significantly affects project costs.

The selection and design of the secondary masses in a TMD system involve careful analysis and consideration of the structural dynamics, encompassing natural frequencies, mode shapes, and damping characteristics, which will be discussed in subsequent chapters.

The objective is to optimize the TMD's effectiveness in reducing vibrations and enhancing the dynamic behavior of the primary structure.

2.2.2 Spring

In TMDs, the mass is typically connected to the primary structure by a spring and a damper. **Coil spring**s are commonly used as the spring component in TMDs due to their versatile characteristics. Coil springs offer several advantageous properties for vibration control, making them well-suited for TMD applications. These properties include:

1. **Customizability:** Coil springs can be designed and manufactured in various sizes, shapes, and configurations to accommodate specific TMD requirements. The number of coils, wire diameter, and coil pitch can be adjusted to achieve the desired spring characteristics.
2. **Wide range of spring constants:** Coil springs provide a broad range of spring constants, allowing for precise tuning of the TMD system. The spring constant determines the stiffness of the spring and influences its ability to resist deformation under load. By selecting an appropriate spring constant, the TMD can effectively address the target vibration frequencies.
3. **High energy storage capacity:** Coil springs have the ability to store a significant amount of mechanical energy when compressed or extended. This energy storage capability enables them to absorb and dissipate vibrations effectively, reducing the amplitude of oscillations in the primary structure.
4. **Damping potential:** While coil springs inherently possess some inherent damping properties, additional damping elements can be incorporated to further enhance the TMD system's performance. These damping elements, such as friction dampers or viscoelastic materials, are often integrated within the coil spring assembly to increase energy dissipation and improve vibration attenuation.
5. **Durability and longevity:** Coil springs are typically made from materials with high strength and durability, such as steel alloys. This allows them to withstand the repeated loading and unloading cycles associated with TMD operation, ensuring long-term reliability and performance.

When designing TMD systems with coil springs, factors such as the mass ratio (the ratio of TMD mass to primary structure mass), spring stiffness, damping characteristics, and natural frequencies of the structure must be considered. These parameters are carefully analyzed and optimized to achieve the desired vibration control objectives.

In conclusion, coil springs offer customizable characteristics, high energy storage capacity, and the potential for damping, making them a valuable component in TMDs. Their versatility, durability, and tunability make coil springs well-suited for effectively reducing structural vibrations and enhancing the dynamic behavior of primary structures.

2.2.3 Damper

For damping component of TMD, **Viscoelastic** or **Eddy current** damping are common to be utilized.

In Viscoelastic damping is a damping mechanism that utilizes the energy dissipation properties of viscoelastic materials to reduce vibrations and dampen mechanical systems. Viscoelastic materials exhibit both viscous (fluid-like) and elastic (solid-like) characteristics, combining the ability to store and release energy over time. Viscoelastic damping typically involves the use of materials such as polymers or rubber compounds that exhibit viscoelastic properties. When subjected to mechanical vibrations, these materials undergo internal friction, causing them to dissipate energy in the form of heat. This energy dissipation reduces the amplitude of vibrations and provides damping effects.

Eddy current damping is a technique used for vibration control and damping in mechanical systems. It involves the generation of eddy currents, which are circular electric currents induced in conductive materials when they are subjected to a changing magnetic field. These eddy currents create a resistive force that opposes the motion of the vibrating system, thus dissipating its kinetic energy and reducing vibration amplitudes.

In the context of structural control systems, by incorporating eddy current dampers within TMDs, the damping characteristics (frequency and damping ratio) of the TMD system can be controlled and maintained consistently, independent of temperature variations. This ensures reliable and effective vibration mitigation in structures subjected to dynamic loads, such as pedestrian-induced vibrations or wind-induced vibrations in bridges.

2.3 Advantages and Shortcomings of TMDs

As we delve deeper into the realm of structural engineering, it becomes increasingly apparent that the stability and safety of structures, particularly tall ones, are governed not just by the strength of materials used, but also by sophisticated control mechanisms designed to mitigate the effects of natural hazards. TMDs are one of such control systems that has has gained significant attention recent years.

This section provides an objective analysis of the benefits and challenges associated with implementing TMDs as a vibration control system in structures. Through this balanced appraisal, engineers, architects, and builders may gain insights into whether the adoption of TMDs aligns with their project goals and constraints. As we will discuss, the decision to implement TMDs involves considerations beyond the initial expense, including the long-term performance, maintenance, and spatial demands of these devices.

Advantages of Tuned Mass Dampers

1. **Simplicity and Dependability:** The inherent advantage of TMDs over active, semi-active, or hybrid control systems lies in their simplicity and reliability. Operating passively, TMDs do not rely on a continuous power supply. This feature increases their dependability, especially in scenarios where power availability is intermittent or unavailable, marking a stark contrast to power-dependent active and hybrid systems.
2. **Lower Operational Costs:** The passive nature of TMDs also implies they have fewer active mechanical parts, leading to reduced operational and maintenance costs compared to their active, semi-active, or hybrid counterparts.
3. **Enhanced Structural Safety:** TMDs enhance the stability and safety of a structure, particularly high-rise buildings, by reducing the structural response to dynamic external loads, such as those induced by wind and earthquakes.
4. **Versatility in Implementation:** TMDs can be integrated into both new and existing structures, providing a valuable option for retrofitting buildings to improve seismic and wind resistance.
5. **Prolonged Structural Life:** By reducing the magnitude and number of loading and unloading cycles on structural components, TMDs can help extending the lifespan of a building.
6. **Improved Comfort for Occupants:** The use of TMDs can dramatically reduce the perceptible sway in high-rise buildings, which can otherwise induce discomfort.

Shortcomings of Tuned Mass Dampers

1. **Limited Adaptability:** One of the main limitations of TMDs is their limited adaptability to changing conditions. Since they are designed to target a specific frequency, with fixed parameters (e.g., mass, stiffness, and damping), TMDs are not inherently capable of adjusting to varying excitation levels or frequencies. As result, their effectiveness diminishes when subjected to vibrations outside their tuned range.
2. **Tuning Challenges:** The effectiveness of a TMD is heavily reliant on it being correctly tuned to the building's natural frequency. Misjudgments can result in ineffective damping or even amplify the vibrations in some cases.
3. **High Initial Costs:** The design, fabrication, and installation of TMDs can be costly. They often require customization to suit the specific dynamics of each building, which can add to the expense.
4. **Space Requirements:** TMDs, especially the larger ones, can consume considerable space within a building, which could otherwise be used for functional purposes.
5. **Maintenance:** TMDs require regular inspection and maintenance to ensure they are functioning correctly. Depending on the type and location of the TMD, this could be difficult and expensive.

In conclusion, while TMDs offer significant advantages in controlling building vibrations and improving structural safety, their implementation also comes with several challenges. A thorough cost-benefit analysis should be conducted before deciding to implement TMDs, taking into consideration the local risk factors such as seismic and wind loads.

2.4 Arrangements of TMDs

TMDs can be implemented in various configurations depending on the specific requirements of the structure and the desired level of vibration control. This section explores three common arrangements: single TMD, multiple TMDs, and distributed systems. Each arrangement provides unique advantages and considerations in the context of vibration mitigation.

1. Single TMD:

A single TMD (STMD) configuration involves the attachment of a single mass-spring-damper system to the primary structure (Fig. 2.8). The TMD is typically positioned at a critical location where it can effectively counteract the dominant vibration mode. This arrangement offers simplicity in design and installation, making it suitable for structures with a single prominent vibration frequency. The previous research and experiences show that, if well tuned, single TMDs can effectively suppress vibrations at the targeted frequency, enhancing the structural performance and reducing the risk of discomfort.

Common application of STMD systems are:

- Bridges
- Towers
- Chimneys
- Shear buildings

These structures typically have a single dominant mode that influences the overall structural response to external dynamic loads. Further discussion of "Dominant modes" and "Dominant frequencies" will be covered in subsequent chapters of this book.

The efficacy of TMDs in countering earthquake-induced vibrations in buildings is well-documented in research. Pioneering work by Den Hartog [104] involved proposing the addition of a mass-spring-damper system to a building to suppress vibrations. His groundbreaking work laid the foundation for determining the optimal TMD parameters for a single degree of freedom (SDOF) structure exposed to harmonic excitations.

Subsequent research has expanded the understanding of single TMDs, exploring everything from analytical formulations to experimental investigations. In terms of design and optimization, studies like those by Arfiadi et al. [33] and Pourzeinali et al. [91] have focused on determining the optimal parameters of single TMDs for maximum vibration reduction.

Fig. 2.8 Single TMD system

Their work, along with others, confirms that properly tuned single TMDs can effectively moderate structural responses.

2. Multiple TMDs:

STMD systems are effective in improving structural responses, particularly during steady-state phases, when the structure exhibits a dominant frequency response. However, many engineering problems require the control of multiple vibration modes. There are several reasons for such cases. For instance, in real buildings, different lateral stiffnesses can result in diverse natural frequencies in different directions. In such scenarios, it becomes advantageous for structural engineers to employ multiple TMD systems to control multiple modes simultaneously. Each TMD is designed and positioned to target specific vibration modes of the structure. By using multiple TMDs, a broader range of vibration frequencies can be addressed, leading to improved overall vibration control.

Multiple Tuned Mass Dampers (MTMDs) were first introduced by Igusa and Xu in 1990 [40] after which they have been studied in several researches [32, 41–45]. The decision to use multiple tuned mass dampers is influenced not only by structural parameters but also by the type of external excitation. Numerous studies have demonstrated the superior performance of multiple tuned mass damper (MTMD) systems in mitigating structural responses under earthquake excitations [26, 34, 44–54]. The optimal placement and parameters of MTMD systems have been extensively investigated in the literature.

Li et al. [55] proposed a novel optimum criterion for optimizing the properties of double TMDs in structures subjected to ground acceleration. Elias et al. [97] investigated the effectiveness of distributed TMD systems in vibration control for chimneys. Chen et al. [46] studied the efficiency of using multiple TMDs in mitigating the seismic responses in a six-story building. After that, Sakr [56] used partial floor loads as MTMDs. These researches show that the MTMD systems cover a wider frequency range and are less sensitive to the uncertainties of the system (Fig. 2.9).

3. Distributed Systems:

Distributed TMD systems involve the installation of multiple masses distributed throughout the structure. Instead of placement of the masses in a single location, they are strategically placed at different points along the structure to target various vibration modes. This arrangement can provide efficient vibration reduction across a wide frequency range, making it suitable for structures with complex vibration characteristics. It is important to note that each arrangement has its own design considerations and limitations. Factors such as mass distribution, tuning of individual TMDs, spatial constraints, and the interaction between

Fig. 2.9 Multiple TMD system

multiple TMDs need to be carefully analyzed during the design phase. Structural dynamics analysis, including mode shapes, natural frequencies, and damping characteristics, plays a crucial role in determining the optimal arrangement of TMDs for effective vibration control.

The selection of a specific TMD arrangement depends on the characteristics of the primary structure, the desired vibration control objectives, and the available resources. This requires structural analyses and optimizations to determine the most appropriate TMD arrangement for a given structure, ensuring effective vibration mitigation and improved structural performance (Fig. 2.10).

2.5 Applications of TMDs

The different application areas of TMDs are presented in this section. For more efficient presentation, the literature is classified based on the structural type so that the reader can more easily find the related inform related to his problem. The goal of this section is to first highlights the main applications of the TMDs to helps to understand in which problems have they already proved their efficiency. The second goal is to familiar the reader with the main previous works on the TMDs which helps the researchers to identify the development potentials.

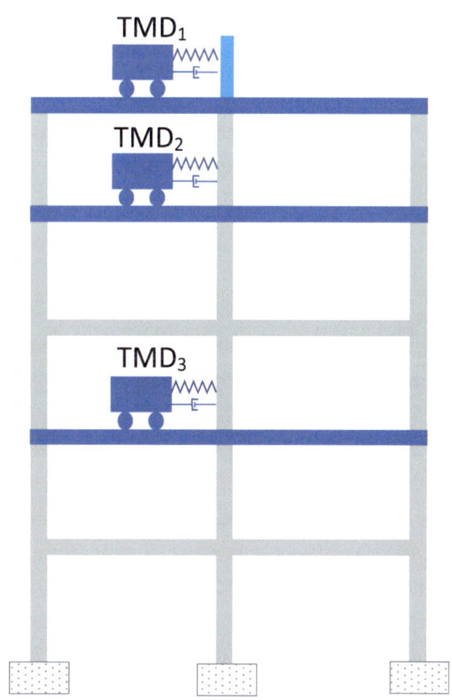

Fig. 2.10 Distributed multiple TMD system

2.5.1 Bridges

Bridges are one of structures that TMDs has a proven efficiency. The following issues are the main areas where TMDs can be utilized as an optimum solution in the bridges:

1. Mitigation of resonance and human-induced vibrations in pedestrian bridges
2. Suppress the vibration in high-speed railway
3. Flutter control in suspension bridges
4. Seismic response control of the bridges
5. Control the wind-induced vibrations
6. Retrofitting the existing bridges

A. Pedestrian Bridges

Utilization of TMDs in pedestrian bridge design serves various purposes, including mitigating resonance effects, reducing lateral displacements under earthquake or wind loads, and improving user comfort by controlling movements. TMDs can be integrated into both new bridge construction projects and retrofitting solutions for existing structures. Two common loading conditions where TMDs prove efficient in reducing structural responses are harmonic loading and random loading, such as white noise. Given the advantages of employing TMDs in bridges, it is crucial for bridge engineers to be familiar with the theories and applications of TMDs, enabling them to consider these systems as an applicable solution for their projects.

The application of TMDs in footbridge design has gained prominence to meet serviceability requirements. Several noteworthy examples highlight the successful implementation of TMDs in footbridge projects. For instance, the Solferino Footbridge in Paris, France [57], the Singapore Changi Airport Footbridge [58], and the Pedro e Inês Footbridge in Coimbra, Portugal [59] demonstrate the efficacy of TMDs in controlling vibrations, ensuring structural integrity, and enhancing user comfort. In another study by Van Nimmen et al. [60], the dynamic performance of a slender steel footbridge was evaluated before and after installing two TMDs, resulting in successful reduction of human-induced vibrations. Li et al. [26] employed multiple TMDs to control resonance caused by crowd-induced vibration in a footbridge. Wang et al. [61] conducted a study on vibration control of a long-span steel pedestrian bridge using TMDs. Carpineto et al. [43] utilized multiple TMDs, including viscoelastic TMDs, to mitigate pedestrian-induced vibrations in a suspension bridge effectively. In another study on reinforced concrete footbridges, Behloul et al. [62] investigated the utilization of multiple TMDs to control vibrations in the Seoul Seonvu footbridge, which spans 120 meters. The study included testing the bridge after TMD installation, demonstrating the efficiency of the system in controlling vibrations.

An exemplary case of utilizing TMDs in retrofitting existing footbridges is the retrofit of the London Millennium Footbridge [2] (Fig. 2.11). This footbridge, opened on June 10, 2000, encountered unexpected lateral movements on its opening day as pedestrians traversed the bridge. Investigations revealed that these movements resulted from lateral forces exerted by pedestrians on the moving deck surface. Importantly, this phenomenon is not exclusive to the bridge's technical innovations and could potentially occur on any bridge with a lateral frequency below approximately 1.3 Hz when a sufficient number of pedestrians are present. The retrofitting goals for the London Millennium Footbridge included reducing movements to acceptable levels, minimizing the maintenance implications of modifications, and preserving the bridge's slim appearance. Ultimately, a combination of TMDs and viscous dampers were designed and implemented to effectively address the issue.

B. Railway Bridges

High-speed railway bridges face significant challenges related to resonance and fatigue under dynamic loads imposed by trains. The resonance between the train loads and the bridge's natural frequency leads to excessive displacements, increasing the risk of accidents and discomfort for passengers. Fatigue further reduces the service life of the bridge and can result in structural element failures over time.

The utilization of TMDs to improve the performance of railway bridges has been the subject of several studies in the literature. Yau [63] investigated the effectiveness of TMDs in suppressing train-induced vibrations in railway bridges.

In another study, Li et al. [64] investigated the application of multiple TMDs (MTMDs) to suppress train-induced vibrations in high-speed railway bridges. Their research focused on optimizing the parameters of MTMDs to mitigate resonance effects. The findings demonstrate the effectiveness of MTMD systems in controlling resonance and bridge movements during dynamic loading. Researchers have proposed various modeling techniques and TMD arrangements for the implementation of multiple TMD systems [41, 65–67]. Luu et al. [68]

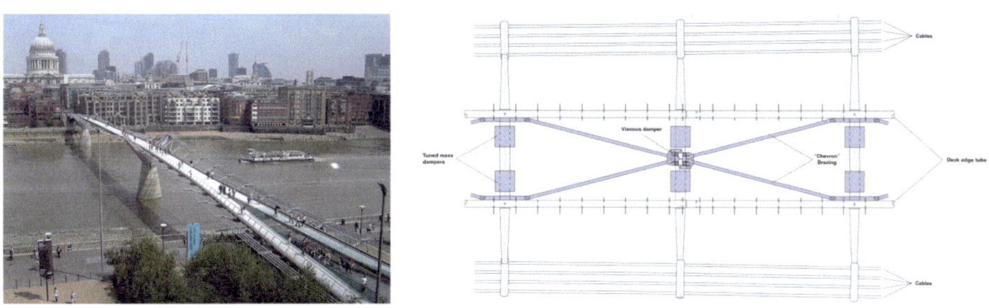

(a) (b)

Fig. 2.11 London millennium footbridge and damper locations [2]

developed a method for direct optimization of MTMDs simultaneously on different modes contributing significantly to the multi-resonant peaks to make the TMDs more effective and robust in suppressing multi-resonant dynamic structural response of high-speed train bridges. Additionally, Andersson et al. [69] investigated the use of TMD systems as a solution to address fatigue issues in structural elements of bridges.

TMDs also are utilized for noise reduction and vibration control of the railways. An example of such TMDs from Wilson Ho and Associates Limited is showed in Fig. 2.12.

These studies highlight the potential of TMDs in mitigating vibrations and enhancing the performance of high-speed railway bridges. The optimization of TMD parameters, utilization of multiple TMD systems, and consideration of fatigue-related concerns contribute to improved safety, passenger comfort, and the extended service life of these critical transportation structures.

C. Flutter Effect

The flutter effect in bridges refers to a phenomenon where the bridge experiences self-excited oscillations due to the interaction between wind and the bridge structure. When wind flows across the bridge, it creates varying pressure distributions that can induce vibrations. If the frequency of the wind-induced vibrations matches or closely approaches the natural frequency of the bridge, resonance occurs, leading to significant oscillations. This can cause detrimental effects such as increased stress levels, fatigue, and potential structural damage. Flutter is a critical consideration in bridge design and requires careful analysis to ensure the bridge can withstand and mitigate these wind-induced vibrations to maintain safe and stable operation.AThe collapse of the Tacoma Narrows Bridge in 1940 stands as a prominent

(a) (b)

Fig. 2.12 Multi directional tuned mass rail damper [3]

example of the catastrophic consequences that can arise from aeroelastic flutter. Under moderate winds, the bridge experienced self-excited and unbounded flutter, leading to its collapse (Fig. 2.13). This incident serves as a stark reminder of the significance of properly accounting for flutter in bridge design and highlighting the importance of implementing appropriate mitigation measures.

Extensive research has been carried out in the literature to assess the effectiveness of TMDs in mitigating flutter phenomena in bridges. Chen et al. [70] studied the optimal parameters of TMDs for controlling flutter effects. In another study, Kwon et al. [71] employed Genetic Algorithm techniques to investigate the optimum parameters of Irregular MTMDs to suppress flutter in suspension bridges. Casalotti et al. [72] implemented hysteric TMDs to mitigate post-flutter oscillations in long-span suspension bridges. Their findings demonstrated that while hysteric TMDs may not significantly alter the flutter condition, they effectively controlled the post-flutter response by reducing the amplitude of limit cycle oscillations to very low levels. Ubertini et al. [73] conducted a parametric study on the reliability-based design of TMDs against bridge flutter. They concluded that a reliability-based design approach can be effectively utilized to enhance robustness against frequency mistuning, reduce costs associated with supplemental damping, and achieve a desired safety performance level for multiple tuned mass damper configurations.

Fig. 2.13 Tacoma narrow bridge failure (1940)

D. Wind Induced Vibrations

Mitigating the movements of bridges during wind and seismic excitations is a critical and challenging task. These movements can give rise to various problems, including local failures in structural elements, loss of support, and potential accidents for bridge users. Consequently, finding effective solutions to address these challenges has remained a prominent research topic.

Regarding seismic excitations, seismic isolation techniques are typically the preferred choice. This is because providing horizontal degrees of freedom is necessary to release temperature strains and also can significantly reduce seismic loads.

The utilization of TMDs for reducing wind-induced vibrations has gained significant attention from researchers. Lin et al. [74] developed an analytical model to assess the performance of MTMDs for long-span bridges subjected to wind excitations. Their findings concluded that MTMDs designed with recommended parameters are more efficient and robust compared to STMDs systems. Sun et al. [75] employed TMDs for vibration control in a long-span bridge with a span length of 944 m. They monitored the performance of the TMDs using a structural health monitoring system during Typhoon Chan-hom, which passed through the area near Chongqi Bridge on July 10–12, 2015. The wind characteristics were calculated using the monitoring data, and the response of the bridge and TMDs were evaluated under the typhoon conditions. The results demonstrated the effectiveness of the TMDs and their functionality under different wind conditions.

In another study, Yin et al. [76] investigated the suppression of wind/traffic/bridge-coupled systems using multiple pounding TMDs. They established coupled equations and studied different arrangements and parameters of the pounding TMDs to achieve optimal performance. Debnath et al. [77] explored multi-modal vibration control of truss bridges using TMDs under general loading. They proposed a strategy based on modal frequency response function on the traditional mode-wise control approach. This proposed approach was applied to the Saraighat Bridge, an important large truss bridge, for the analytical design of a TMD system considering general loading conditions. Their results demonstrated good performance of the designed TMD system under various white-noise-based general loading conditions. Niu et al. [78] investigated the effectiveness of TMDs with eddy current damping in wind-induced vibrations. They proposed a new eddy current damping system for large-scale civil infrastructure applications, utilizing permanent magnets to eliminate the need for power and a combination of copper and steel plates to improve energy dissipation efficiency.

E. Retrofitting the Existing Bridges

TMDs are also known as an effective retrofitting solution for bridges. Retrofitting bridges with TMDs can address various structural challenges and enhance their performance under dynamic loading conditions. There are several reasons why TMDs are considered a favorable retrofitting solution for bridges:

1. **Vibration Control:** One of the primary reasons for utilizing TMDs as a retrofitting solution is their ability to control vibrations. Over time, bridges may experience increased vibrations due to changes in usage patterns, increased traffic volume, or aging of structural elements. TMDs can effectively mitigate these vibrations, reducing structural stresses and enhancing the comfort and safety of bridge users.
2. **Cost-Effective Solution:** Retrofitting existing bridges with TMDs can be more cost-effective than undertaking extensive structural modifications or reconstructions. TMDs can provide targeted vibration control without the need for major alterations to the existing bridge structure.
3. **Compatibility withVarious Bridge Types:** TMDs offer versatility in their application to different types of bridges. Whether it is a suspension bridge or truss bridge, TMDs can be customized and installed to suit the specific structural characteristics of the bridge. This adaptability allows for widespread utilization of TMDs as retrofitting solutions across a wide range of bridge designs.
4. **Non-Intrusive Installation:** Retrofitting bridges with TMDs typically involves non-intrusive installation procedures. The TMDs can be strategically placed at appropriate locations within the bridge structure, such as the deck, piers, or cables, to effectively control vibrations. This non-intrusive nature of TMD retrofitting minimizes disruptions to bridge operations during the installation process.
5. **Improved Service Life:** By reducing the dynamic response of the bridge, TMDs can help extend the service life of aging bridge structures. Vibrations, especially those induced by wind or traffic, can accelerate fatigue and contribute to the deterioration of bridge components. The implementation of TMDs as retrofitting solutions can mitigate these vibrations, reducing fatigue damage and improving the longevity of the bridge.

In the case that the structural engineer decides to utilize TMDs as a retrofit solution for a bridge, several key considerations should be kept in mind to ensure a successful implementation such as:

1. Structural Assessment: The engineer should conduct a thorough structural assessment of the bridge to identify the specific vibration issues and determine whether TMDs are the appropriate retrofitting solution. This assessment involves evaluating the bridge's dynamic characteristics, natural frequencies, and response under different loading conditions.
2. Retrofit Objectives: Clear retrofit objectives should be established to guide the design and installation of the TMD system. The engineer needs to quantify objectives such as reduction in vibrations, improve structural integrity, enhance comfort for bridge users, or achieve specific performance criteria.

3. Dynamic Analysis: Appropriate dynamic analysis should be done for designing TMDs. A Mathematical model of the bridge should be created to simulate the dynamic behavior of the controlled sysyem under various loading conditions, such as wind or traffic-induced vibrations. This analysis helps determine the optimal TMD parameters, including mass, stiffness, and damping coefficients based on retrofitting goal.
4. TMD Configuration and Placement: The engineer must carefully consider the TMD configuration and placement within the bridge structure. Factors such as the number of TMDs, their size, and their locations should be determined to achieve the structural behaviour reduction while considering practical and installation constraints. The TMDs should be strategically positioned to effectively counteract the bridge's dynamic responses.
5. Structural Compatibility: The TMD retrofit solution should be compatible with the existing bridge structure. The engineer should ensure that the bridge can accommodate the additional weight, forces, and movements introduced by the TMD system. Structural integrity and stability should not be compromised, and any necessary strengthening or modifications should be considered.
6. Monitoring and Maintenance: The engineer should incorporate a monitoring system to assess the performance of the TMD retrofit and ensure its long-term effectiveness. Regular maintenance and inspection protocols should be established to monitor the condition of the TMD components, including the damping mechanisms and connection points, to ensure optimal functionality and safety.
7. Collaborative Approach: It is essential for the structural engineer to collaborate with TMD manufacturing companies. Their experience can provide valuable insights and assist in the design, analysis, and implementation of the TMD retrofit solution.

2.5.2 Buildings

The utilization of TMDs in buildings has garnered significant attention as an effective solution for mitigating both seismic and wind-induced vibrations. Extensive research has been conducted to enhance the performance of TMD systems in vibration control for buildings.

Initially, studies primarily focused on STMD systems to evaluate their efficiency in vibration control for buildings. Early works by Sladek et al. [79], McNamara et al. [80], Sadek et al. [81], and Luft et al. [82] evaluated the efficiency of STMDs in controlling vibrations. Kaynia et al. [83] conducted a time-history analysis on a single degree of freedom system to demonstrate the effectiveness of STMD systems in seismic control.

Further research explored the application of TMDs in various scenarios. Liu et al. [84] utilized TMDs to improve structural responses in a high-rise building considering soil-structure interactions. Aly [85] proposed robust TMDs for response control in buildings under multi-directional wind loads. Arfiadi et al. [33] employed a hybrid genetic algorithm to determine the optimum properties and location of a TMD for a 10-story building under earthquake excitation.

It is important to note that TMD performance under earthquake excitations is also related to the characteristics of the ground motion. TMDs are shown to be more effective for structures subjected to narrow band long-duration far-fault (FF) excitations compared to pulse-like near-fault (NF) ground motions. Single TMD systems are sensitive to mistuning and uncertainties, leading to the introduction of Multiple Tuned Mass Dampers (MTMDs) as a solution. Igusa and Xu [40] introduced MTMDs, which have since been the subject of numerous studies by researchers such as Sakr et al. [56], Zuo et al. [86], Elias et al. [34], Hoang et al. [45], Jokic et al. [48], Li et al. [55], and Pisal et al. [87]. Li et al. [88] studied the optimum design parameters for TMDs in multiple TMD configurations using a SDOF system. Chen et al. [46] investigated the efficiency of multiple TMDs in mitigating seismic responses in a six-story building, while Sakr [56] utilized partial floor loads as MTMDs. These studies revealed that MTMD systems cover a wider frequency range and are less sensitive to uncertainties compared to single TMD systems.

Notable tall buildings have also benefitted from TMD utilization to control vibrations (see Fig. 2.14). Most of these buildings are equipped with a single TMD placed at the top level, as the top levels experience maximum modal displacement due to wind excitation.

Studies have shown that placing a single TMD with a tuning frequency close to the fundamental structural frequency effectively reduces wind-induced motions [89, 90]. Elias et al. [34] explored the use of distributed MTMD systems in reducing wind-induced vibrations,

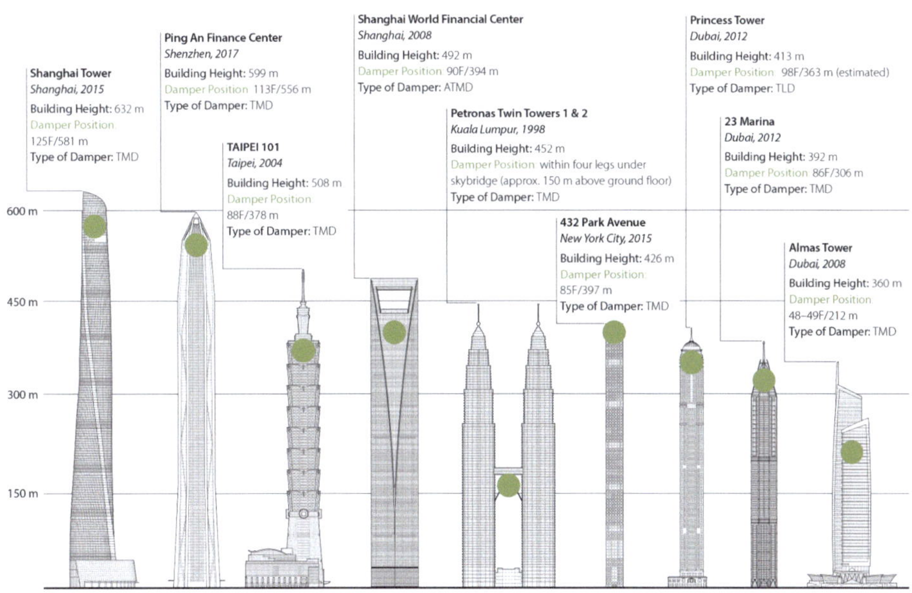

Fig. 2.14 Tallest completed buildings with dampers [4]

finding them more effective than single TMD systems or MTMD systems with all TMDs placed at the top level.

TMDs have also proven effective in mitigating earthquake vibrations in buildings. Pourzeinali et al. [91] employed multi-objective optimization to determine the design parameters of a TMD for a 12-story building under earthquake excitation. Li [55] proposed an optimum criterion for optimizing the properties of double TMDs for structures subjected to ground acceleration. Elias et al. [92] studied the effectiveness of distributed TMD systems in vibration control of chimneys.

The existing research on tall buildings has predominantly focused on studying TMD parameters under wind-induced vibrations or investigating the effectiveness of a single TMD under earthquake excitation. However, studying the optimal parameters of TMDs under earthquake excitation, without limiting the number and location of the TMDs, remains a challenging task. This difficulty arises due to the stochastic nature of earthquakes and the complex seismic behavior exhibited by tall buildings, necessitating extensive and thorough studies.

Moreover, unlike wind loads, during earthquakes, the higher modes of vibration tend to have a more significant contribution to the overall response of tall buildings. This is primarily due to two reasons. First, tall buildings exhibit lower frequencies for their higher modes compared to low- and mid-rise structures. Second, earthquakes have a wide frequency content that can activate multiple modes in tall buildings. Consequently, solely controlling the lower modes by placing TMDs at the top levels may not lead to an optimal solution for effectively controlling the motions of tall buildings during earthquakes.

3 Applicable Theories

3.1 Introduction

In the previous chapter, we reviewed the literature on the application of TMDs across various structural types. To ensure the successful implementation of TMD systems, it is crucial to have a solid understanding of the underlying theories and methodologies involved in modeling, analysis, design, optimization, and evaluation of structures equipped with TMDs.

While there are numerous reference books available on structural dynamics and tuned mass dampers, navigating through vast amounts of information can be time-consuming and lead to confusion. Therefore, this chapter aims to streamline the process by presenting selected and useful theories that are specifically relevant to TMD applications. By providing concise and practical insights into the key theories, this chapter will save time, enhance understanding, and facilitate informed decision-making during the implementation of TMD systems.

The following sections will be covered in this chapter:

1. Equations of Motion: An overview of the fundamental equations that govern the motion of structures, providing the necessary foundation for analyzing TMD-equipped systems.
2. State-Space Model: The state-space representation, a powerful tool for describing the dynamic behavior of structures with TMDs, enabling comprehensive system analysis and control design.
3. Newmark-β Method: An efficient numerical integration scheme commonly used for solving dynamic equations of motion, enabling accurate and efficient simulations of TMD-equipped structures.
4. Static Condensation: A technique for reducing the dimensionality of the equations of motion, simplifying the analysis of large-scale structures with TMDs.

H. Radmard Rahmani, *Seismic Control of Structures Using Tuned Mass Dampers*,
Synthesis Lectures on Engineering, Science, and Technology,
https://doi.org/10.1007/978-3-031-71217-3_3

5. Rayleigh Damping: An empirical damping model widely used in structural dynamics to capture energy dissipation characteristics, offering insights into the damping determination based on other structural parameter.
6. Modal Analysis: A powerful technique for extracting the natural frequencies, mode shapes, and modal damping ratios of structures, aiding in evaluation of the structure and providing useful information for TMD design.
7. Checking the Structure for Resonance Issues: Methods for identifying resonance conditions in the structures.
8. Ground Motion Selection: Strategies for selecting appropriate ground motion records for realistic dynamic analysis, considering the influence of different seismic events on TMD-equipped structures.
9. Frequency Domain vs Time Domain: A comparison between frequency domain and time domain analysis approaches, discussing their advantages and limitations in TMD system evaluation.
10. Genetic Algorithm: An optimization algorithm widely used in TMD design and parameter selection, offering a systematic approach to achieving optimal performance.

By delving into these key theories and methodologies, this chapter equips researchers and engineers with the necessary tools to model, analyze, design, optimize, and evaluate structures equipped with TMDs.

3.2 Equation of Motion

The equation of motion is a fundamental concept in structural engineering that describes the dynamic behavior of a structure. It represents the relationship between the forces acting on a structure and its resulting motion. The equation of the motion for a structure which is equipped with TMDs is as follow:

$$[M]\{\ddot{U}\}+[C]\{\dot{U}\}+[K]\{U\}=\{P_t\} \tag{3.1}$$

where, M, K, and C represent the mass, the stiffness and the damping matrices of the structure and the TMDs:

$$[M]=[M_{st}]+[M_t] \tag{3.2}$$

$$[C]=[C_{st}]+[C_t] \tag{3.3}$$

$$[K]=[K_{st}]+[K_t] \tag{3.4}$$

Indexes *st* and *t* indicate the Degrees of Freedom (DOFs) of the building and the TMDs, respectively.

In case of seismic excitations, the external load vector, P_t, in can be defined as follows:

$$\{P_t\} = -[M]\{1_t\}\ddot{u}_g \tag{3.5}$$

where $\{1_t\}_{(N+n)\times 1} = [1\ 1\ \cdots\ 1]^T$ and the term $\ddot{u}_g$ represents the ground accelerations.

The structural responses, including displacement, velocity, and acceleration matrices, can be expressed as follows:

$$\{U\} = \{u_{st1}, u_{st2}, \ldots, u_{stN}, u_{t1}, u_{t2}, \ldots, u_{tn}\} \tag{3.6}$$

$$\{V\} = \{v_{st1}, v_{st2}, \ldots, v_{stN}, v_{t1}, v_{t2}, \ldots, v_{tn}\} \tag{3.7}$$

$$\{A\} = \{a_{st1}, a_{st2}, \ldots, a_{stN}, a_{t1}, a_{t2}, \ldots, a_{tn}\} \tag{3.8}$$

In these equations, N and n represent the number of degrees of freedom (DOF) for the building and the TMDs respectively. Therefore, the dimensions of the M, K, and C matrices are $(N + n) \times (N + n)$.

In the equation of the motion, the displacements, velocities, accelerations, and the load are functions of the time.

In order to solve the equation of motion for the structures equipped with TMDs, numerical methods are commonly employed. These methods provide efficient and accurate solutions to the complex dynamic behavior of structures subjected to external forces and vibrations. One widely used numerical method is the Newmark-β method, which is a time integration scheme that enables the numerical approximation of the equations of motion. This method is widely utilized in the research as it allows for the efficient simulation of the dynamic response of TMD-equipped structures by dividing time into discrete intervals and iteratively solving for the displacements and velocities of the structure.

Another numerical technique is the finite element method (FEM), which discretizes the structure into smaller elements to approximate the behavior of the system. FEM is particularly useful for analyzing complex structures with TMDs, as it allows for the consideration of geometric nonlinearity and material properties. Additionally, other numerical methods such as the modal superposition method and the direct integration method can also be employed to solve the equations of motion for TMD-equipped structures. The numerical methods provide engineers with powerful tools to accurately predict the response of structures and optimize the design of TMD systems for effective vibration control.Therefore, in order to obtain the structural responses, the equation of the motion shall be solved in each time step.

3.3 State-Space Model

In structural control problems, it is common to utilize the *state-space* representation of a structure instead of relying solely on the equations of motion discussed earlier. The equation of the motion is a second order differential equation. In the state-space model, the equation of the motion will be transformed into two first order differential equations.

The state-space representation offers several advantages, particularly in the context of control problems, as it allows for more effective manipulation of system parameters. In this representation, a *state* vector is defined, which encapsulates the current response of the structure. The state vector consists of various state variables, such as displacements, velocities, or other relevant quantities, which collectively describe the dynamic behavior of the system. By formulating the problem in the state-space framework, engineers can employ various control techniques and algorithms to analyze and optimize the behavior of structures equipped with TMDs. This representation enables a comprehensive understanding of the system's response and facilitates the design of control strategies. In follow, transforming the equation of motion to the state-space model is explained.

In order to build the state-space model, we generally consider the vector $X = \{\mathrm{u},\ \dot{\mathrm{u}}\}^T$ as state vector which includes displacement vector and the derivative of the displacements with respect to time which represents the velocity vector. We also define a new variable for the velocity as follow:

$$\nu = \dot{u} \tag{3.9}$$

Substituting in the Eq. 3.1:

$$[m]\{\dot{\nu}\} + [c]\{\nu\} + [k]\{u\} = p$$

In this equation, p represents the external loads. In case of analysis under earthquake loads:

$$p = -[m]\{I\}\ddot{u}_g \tag{3.10}$$

By sorting both equation we can build a system of the equations:

$$\left\{\begin{array}{c} \nu = \dot{u} \\ \{\dot{\nu}\} = -[m]^{-1}[c]\{\dot{u}\} - [m]^{-1}[k]\{u\} + [m]^{-1}p \end{array}\right\} \tag{3.11}$$

As we defined, the state vector X is :

$$X = \left\{\begin{array}{c} u \\ \dot{u} \end{array}\right\} \tag{3.12}$$

The derivative of X with respect to time:

$$\dot{X} = \begin{Bmatrix} \dot{u} \\ \ddot{u} \end{Bmatrix} = \begin{Bmatrix} \nu \\ \dot{\nu} \end{Bmatrix} \tag{3.13}$$

The Eq. 3.11 can be written as follow:

$$\dot{X} = AX + Bp \tag{3.14}$$

Where:

$$A = \begin{bmatrix} 0 & 1 \\ -\frac{k}{m} & -\frac{c}{m} \end{bmatrix} \tag{3.15}$$

$$B = \begin{bmatrix} 0 \\ m^{-1} \end{bmatrix} \tag{3.16}$$

The state-space equation can be solved using numerical methods.

We also define an output vector Y which represents the measured or observed quantities of interest, C is the output matrix that relates the state variables to the outputs, and D is the feedforward matrix that relates the control inputs directly to the outputs. For example:

$$Y = CX + Dp \tag{3.17}$$

These coefficients can be determined based on the specific objectives of the control system and the measurements or observations required. For example, if the goal is to monitor and control the displacements of the structure and the TMD displacement, the output matrix C can be selected accordingly. For example:

$$C = \begin{bmatrix} 1\,0 \\ 0\,1 \end{bmatrix} \tag{3.18}$$

Similarly, the feedforward matrix D can be defined based on the desired relationship between the control inputs and the outputs.

3.4 Degrees of Freedom Reduction Methods

In the domain of structural engineering, accurately modeling and analyzing large, complex structures can often lead to a finite element model with a substantial number of degrees of freedom (DOF). Handling such a large number of DOF can make the computational process both time-consuming and resource-intensive. Thus, it becomes vital to employ techniques that can reduce the number of DOF without significantly compromising the accuracy of the model's response. This is where DOF reduction methods play an essential role.

This section delves into three prominent methods for reducing the degrees of freedom in a finite element model of a structure, allowing for a more efficient and tractable analysis. These methods are:

1. **Static Condensation:** This method leverages the partitioning of DOFs into primary and secondary sets, subsequently eliminating the secondary set by expressing them in terms of the primary DOFs. It is particularly effective when certain parts of the structure are considerably stiffer or when focusing on specific behavior.
2. **Guyan Reduction:** Named after its developer, Robert J. Guyan, this method also focuses on partitioning the DOFs but simplifies the process by neglecting higher-frequency contributions. The Guyan reduction technique can be especially useful when interested in the structure's behavior at lower frequency ranges.
3. **Ritz Transformation Method:** The Ritz method is a more flexible approach, allowing the engineer to select specific functions, known as Ritz functions, that represent the primary behavior of the structure. These functions are usually tailored to the specific problem at hand, satisfying boundary conditions and incorporating insights about the structure's essential behavior. Each of these methods offers distinct advantages and applications, depending on the particular structure and the analytical requirements. Proper understanding and application of these techniques can lead to substantial savings in computational time and resources, facilitating the analysis of even highly complex structures.

In the following sections, we will explore each of these methods in detail, providing insights into their underlying principles, implementation, and areas of applicability.

3.4.1 Static Condensation

The static condensation technique is a widely used method in research to reduce the size of structural matrices without compromising accuracy. The computational costs in dynamic analysis are heavily influenced by the size of the structural matrices, as the equation of motion must be solved for each time step. The main idea behind static condensation is to eliminate DOFs with zero mass. Therefore, considering the equation of the motion excluding the damping:

$$[M]\left\{\ddot{U}\right\}+[K]\left\{\dot{U}\right\}=\{P_t\} \tag{3.19}$$

we now classify the DOFs of the structure into the DOFs with mass and without mass.

$$\begin{bmatrix} m_{tt} & 0 \\ 0 & 0 \end{bmatrix}\begin{Bmatrix} \ddot{u}_t \\ \ddot{u}_0 \end{Bmatrix}+\begin{bmatrix} k_{tt} & k_{t0} \\ k_{0t} & k_{00} \end{bmatrix}\begin{Bmatrix} u_t \\ u_0 \end{Bmatrix}=\begin{Bmatrix} P_t(t) \\ 0 \end{Bmatrix} \tag{3.20}$$

where:

u_t: Displacement vector for DOFs with mass
u_0: Displacement vector for DOFs without mass

and $k_{t0}=K_{0t}^T$

Equation 3.20 yields the following equations:

$$m_{tt} \cdot \ddot{u}_t + k_{tt} \cdot u_t + k_{t0} \cdot u_0 = P_t(t) \tag{3.21}$$

$$k_{0t} \cdot u_t + k_{00} \cdot u_0 = \mathbf{0} \tag{3.22}$$

By solving the Eq. 3.22, we can obtain the relationship between the dynamic DOFs, u_t , and DOFs with zero mass, u_0:

$$u_0 = -k_{00}^{-1} \cdot k_{0t} \cdot u_t \tag{3.23}$$

Substituting u_0 in (3.21):

$$m_{tt} \cdot \ddot{u}_t + k_{tt} \cdot u_t + k_{t0} \cdot (-k_{00}^{-1}) \cdot k_{0t} \cdot u_t = P_t(t) \tag{3.24}$$

Which can be written as:

$$m_{tt} \cdot \ddot{u}_t + \left(k_{tt} - k_{0t}^T \cdot k_{00}^{-1} \cdot k_{0t}\right) \cdot u_t = P_t(t) \tag{3.25}$$

In Eq. 3.25, only the dynamic DOFs are present. We can define a condensed stiffness matrix as:

$$\hat{k}_{tt} = k_{tt} - k_{0t}^T \cdot k_{00}^{-1} \cdot k_{0t} \tag{3.26}$$

And Eq. 3.25 can be then rewritten as:

$$m_{tt} \cdot \ddot{u}_t + \hat{k}_{tt} \cdot u_t = P_t(t) \tag{3.27}$$

To provide a summary of the condensation process for mass and stiffness matrices, the following steps are typically followed:

1. Identify the dynamic DOFs, which are usually associated with horizontal displacements.
2. Sorting the DOFs so that the mass and the stiffness matrices be divided into four parts and shown in Fig. 3.1.
3. Separate the mass, stiffness, and force matrices into two parts based on the dynamic DOFs.
4. Calculate the condensed stiffness matrix using the Eq. 3.26.
5. Determine the damping matrix using the Rayleigh damping concept, as described in Sect. 3.5.
6. Apply the condensed matrices in structural analysis using methods such as the Newmark-β method as as discussed in Sect. 3.6

As an example, the stiffness matrix of a frame is presented in Fig. 3.1. The frame has a total of six DOFs, with three DOFs at each node. The DOFs 1 and 4 are associated with horizontal displacements, which are considered dynamic DOFs. To obtain the condensed stiffness matrix, the DOFs are sorted so that the dynamic DOFs are placed first, followed by the other DOFs. The resulting stiffness matrix is shown in Fig. 3.2. Using Eq. 3.26, the condensed stiffness matrix can be computed. Having the condense stiffness matrix, and rearranging the mass matrix, the equation of motion can be formulated using the condensed structural matrices, as indicated in Eq. 3.27.

The concept of static condensation is further demonstrated in the examples presented in Chap. 5, providing practical applications and insights into its implementation.

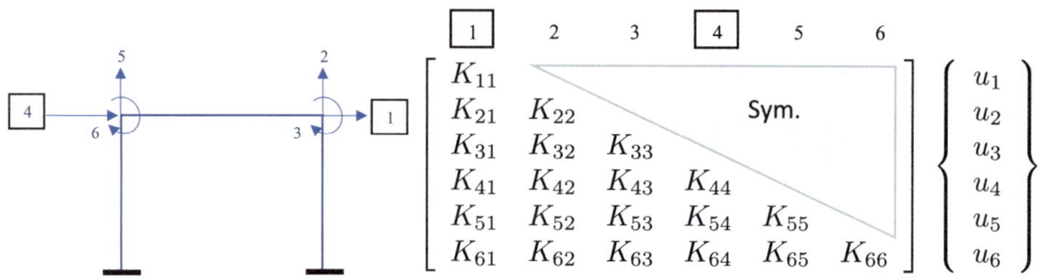

Fig. 3.1 Stiffness matrix of a moment frame

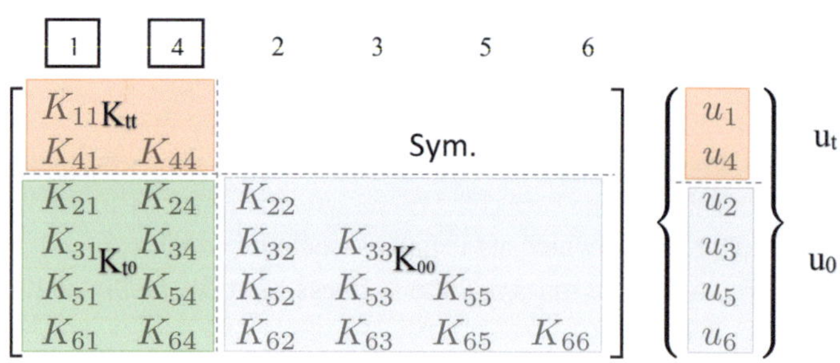

Fig. 3.2 Sorted stiffness matrix of the moment frame

3.4.2 Guyan Reduction Method

The Guyan Reduction Method is a technique used to reduce the degrees of freedom in a finite element model, particularly when dealing with dynamic analyses such as earthquake excitation. It's a static reduction method that allows engineers to simplify complex models, making the computational process more manageable (Fig. 3.3).

Here's how the Guyan Reduction Method works:

1. **Partition the Stiffness Matrix:** The stiffness matrix [K] of the system is partitioned into two submatrices. The first part, $[K_{rr}]$, corresponds to the retained (or master) DOF, and the second part, $[K_{rc}]$, corresponds to the condensed (or slave) degrees of freedom.
2. **Identify Retained and Condensed DOFs:** Select the DOFs to retain in the reduced system, usually those associated with critical components or behavior. The remaining DOFs will be condensed out. The retained DOFs are typically chosen based on the primary response characteristics of the structure, such as critical load paths, while the condensed DOFs often represent secondary or less critical structural characteristics.
3. **Formulate Transformation Matrix:** A transformation matrix [T] is formulated that relates the retained DOFs to the condensed DOFs. This matrix is used to reduce the size of the stiffness and mass matrices. This matrix is created based on the following relation:
$$[K_{rc}] = [K_{rr}]^{-1}[K_{cr}] \tag{3.28}$$
4. **Compute Reduced Matrices:** Utilizing the transformation matrix $[T]$, the reduced stiffness matrix $[K_r]$ and reduced mass matrix $[M_r]$ are computed:
$$[K_r] = [T]^T[K][T][M_r] = [T]^T[M][T] \tag{3.29}$$
5. **Reduce External Forces:** If external forces are involved, they can be reduced as well by multiplying the transformation matrix $[T]$ with the force vector:
$$[F_r] = [T]^T[F] \tag{3.30}$$
6. **Solve Reduced System:** With the reduced matrices in place, the equations of motion can be solved for the retained DOFs, considerably reducing the computational time required.
7. **Recover Original DOFs :** If necessary, the condensed DOFs can be recovered using the transformation matrix:

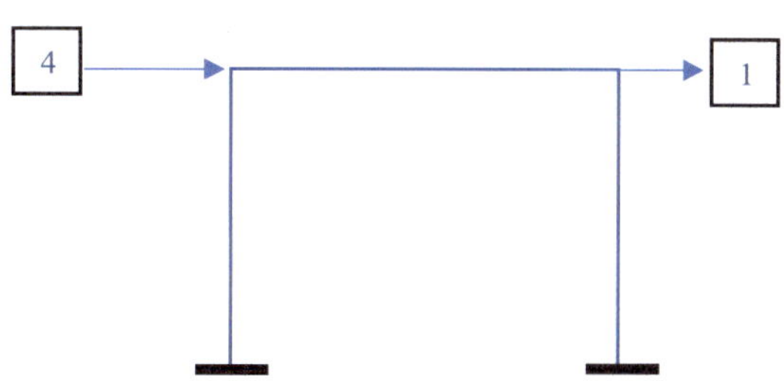

Fig. 3.3 Moment frame with only dynamic DOFs

$$[u_c] = [T][u_r] \tag{3.31}$$

It's worth noting that while the Guyan Reduction Method can significantly reduce computational complexity, it's based on a static approximation and might not accurately capture all dynamic characteristics. Its efficiency depends heavily on the choice of retained DOFs. Other reduction techniques, such as modal truncation, may be combined with Guyan Reduction to enhance accuracy in specific applications.

3.4.3 Ritz Transformation Method

The Ritz Transformation Method is a powerful approach used in finite element analysis for reducing the DOFs in a structure, without losing critical information regarding its behavior. This method is particularly relevant in cases where dynamic response is of interest. Here's an overview of the essential steps involved in the Ritz Transformation Method:

1. **Selection of Ritz Functions:** The choice of appropriate Ritz functions is crucial, as they must closely represent the primary behavior of the structure. These functions must satisfy the structure's boundary conditions and are typically chosen based on an engineer's judgment, experience, and deep understanding of the structural behavior. Selection requires a careful examination of the problem's physical characteristics and intuition about how the structure may respond. For example, in a structure with a considerably stiff deck, the deck may be assumed as rigid. This assumption would imply that all the horizontal displacements of the nodes on each floor are the same. Such a simplification is derived from understanding the rigidity of the deck and how it impacts the overall movement, and it helps to reduce complexity without losing critical information about the structural response.
2. **Formulate Ritz Vectors:** Construct the Ritz vectors by combining the chosen Ritz functions. These vectors create a transformation matrix that can be used to reduce the DOFs in the model.
3. **Compute Generalized Mass Matrix:** Multiply the transpose of the Ritz vectors with the original mass matrix, and then with the Ritz vectors again. This will form a generalized mass matrix that represents the system's inertial properties in the reduced space.

$$[M_r] = [R]^T[M][R] \tag{3.32}$$

4. **Compute Generalized Stiffness Matrix:** Similarly, compute the generalized stiffness matrix by multiplying the transpose of the Ritz vectors with the original stiffness matrix, followed by the Ritz vectors. -

$$[K_r] = [R]^T [K][R] \tag{3.33}$$

5. **Transform External Forces:** If the system is subjected to external forces, these must be transformed into the reduced space using the Ritz vectors.

$$[F_r] = [R]^T [F] \tag{3.34}$$

6. **Solve Reduced System of Equations:** The reduced system of equations can be solved using standard techniques. Since the matrices are smaller, the computational cost will be lower.
7. **Recover Full Solution:** Use the Ritz vectors to transform the reduced solution back into the original space, if the full solution is required.

Advantages and Limitations

The Ritz Transformation Method provides substantial computational efficiency, particularly for systems with a large number of DOFs. Its accuracy depends on the choice of Ritz functions, and it can capture complex dynamic behavior if the functions are chosen wisely.

However, the method also has limitations. It requires a deep understanding of the structural behavior to select appropriate Ritz functions. A poor choice can lead to a loss of critical information, especially related to higher mode shapes that may be significant in dynamic analysis.

Application in Earthquake Engineering

In earthquake engineering, where large and complex structures are analyzed under seismic excitations, the Ritz Transformation Method offers an effective way to reduce computational effort. It allows for the consideration of significant dynamic characteristics while reducing the size and complexity of the problem.

By employing the Ritz Transformation Method in the modeling process, engineers can make informed decisions about the structural design and performance under earthquake loads, balancing computational efficiency with the need to accurately represent critical structural behavior.

3.5 Rayleigh Damping

Rayleigh damping is a commonly used approach for incorporating damping in structural analysis in both research and practice. This method allows for the distribution of damping throughout the structure, making it suitable for homogeneous structures where the material and structural systems remain consistent along the height. However, in the case of composite structures (such as those combining steel and concrete), alternative forms of damping should be considered for accurate structural analysis.

In Rayleigh damping, the damping matrix is calculated by combining the factored mass and stiffness matrices. The mass-proportional and stiffness-proportional matrices are defined as follows:

$$C_1 = \alpha \cdot M \tag{3.35}$$

$$C_2 = \beta \cdot K \tag{3.36}$$

The total damping of the structure is considered as sum of the mass-proportional and stiffness-proportional matrices:

$$C = C_1 + C_2 \tag{3.37}$$

As the damping plays an important role in dynamic response of the structure, understanding the concept of Rayleigh damping is crucial for determining structural behavior under dynamic loads. The value for α and β are determining the relative contribution of mass and stiffness to the overall damping behavior. To grasp this concept more effectively, let's consider two extreme cases: when α is zero indicating that the damping is independent of the mass of the structure, and when β is zero indicating that the damping is independent of the stiffness of the structure. By varying the values of α and β, we can tailor the damping characteristics to better match the specific dynamics and behavior of the structure under consideration.

As in structural engineering, it is often convenient to assume a value for the Lehr's damping (ζ) based on the structural specifications, it will be also practical to know to to determine the constants α and β based on the value of Lehr's damping. In this case, we calculate the values of α and β based on the value of ζ for a particular mode n.

The damping ratio is defined as follow:

$$\zeta = \frac{c}{2m\omega_n} = \frac{c\omega_n}{2k} \tag{3.38}$$

By applying the Rayleigh damping concept, the Lehr's damping can be divided into two components: one that is dependent on the mass of the structure and another that is dependent on the stiffness. The total damping value is obtained by summing these two components.

The Lehr's damping for nth mode will be:

$$\zeta_n = \zeta_{1n} + \zeta_{2n} \tag{3.39}$$

Where ζ_{1n} and ζ_{2n} are mass dependent and stiffness dependent components.

The mass-dependent damping component for the nth mode is:

$$C_{1n} = \alpha \cdot M_n \tag{3.40}$$

Therefore:

$$\zeta_{1n} = \frac{C_{1n}}{2M_n\omega_n} = \frac{\alpha}{2\omega_n} \tag{3.41}$$

The stiffness-dependent damping component for the nth mode is:

$$C_{2n} = \beta \cdot K_n \tag{3.42}$$

Similarly, having ζ_n and ω for the mode j, the factor β can be determined as follow:

$$\zeta_{2n} = \frac{C_{2n}\omega_n}{2K_n} = \frac{\beta\omega_n}{2} \tag{3.43}$$

Based on Eq. 3.39, by summing the mass and stiffness components for the nth mode, the total damping ratio for the nth mode is as follow:

$$\zeta_n = \frac{\alpha}{2}\frac{1}{\omega_n} + \frac{\beta}{2}\omega_n \tag{3.44}$$

By writing Eq. 3.44 for two natural modes of the structures, the values for α and β can be calculated as follow:

$$\frac{1}{2}\begin{bmatrix} \frac{1}{\omega_i} & \omega_i \\ \frac{1}{\omega_j} & \omega_j \end{bmatrix}\begin{Bmatrix} \alpha \\ \beta \end{Bmatrix} = \begin{Bmatrix} \zeta_i \\ \zeta_j \end{Bmatrix} \tag{3.45}$$

$$\alpha = \zeta\frac{2\omega_i\omega_j}{\omega_i + \omega_j} \tag{3.46}$$

$$\beta = \zeta\frac{2}{\omega_i + \omega_j} \tag{3.47}$$

Having the values of α and β, the damping matrix C can be obtained for any mode using Eq. 3.37.

A crucial aspect is accurately determining the values of Lehr's damping for two specific modes. In practical terms, a common approach for typical structures is to select the two modes with the highest mass participation ratios in the x and y directions. These modes play a significant role in the overall dynamic behavior of the structure, making them suitable candidates for determining the appropriate values of ζ.

A Matlab code for calculating the damping matrix of the structure using Rayleigh damping is presented in Appendix A.

For steel structures, typical values of Lehr's damping range between 0.01 and 0.03. In the case of concrete structures this value often fall within the range of 0.03 to 0.05. Concrete exhibits higher inherent damping compared to steel due to its viscoelastic nature.

3.6 Newmark-β Method

Newmark method is one of the numerical methods for solving the equation of the motion. In these methods, the structural responses in each time-step will be utilized to determine the responses in the next time-step. The idea behind the Newmark method is to correlate the velocity in the next time-step, $t_i + 1$, to its value in time step t_i, and the accelerations in the time-steps i and $i + 1$ as follow:

$$\dot{u}_{i+1} = \dot{u}_i + \Delta t \cdot (1 - \gamma)\ddot{u}_i + \Delta t \cdot \gamma \ddot{u}_{i+1} \tag{3.48}$$

where: $0 \leq \gamma \leq 1$

The γ factor determines the weight of acceleration in time-steps i and $i + 1$. For example considering a value of 0.5 for the γ value means that we consider a mean value of accelerations in time steps t_i and t_{t+1} . Having the velocity, the displacement response in each time-step is determined as follows:

$$u_i = u_i + \Delta t \cdot \dot{u}_i + (\Delta t)^2 \cdot (0.5 - \beta)\ddot{u}_i + (\Delta t)^2 \cdot \beta \ddot{u}_{i+1} \tag{3.49}$$

where $\frac{1}{4} \leq \beta \leq \frac{1}{2}$

By considering the values of $\gamma = \frac{1}{2}$ and $\beta = \frac{1}{4}$, the mean of the accelerations in two following time-steps will be utilized for calculating the velocity and displacements.

Another special case is the consideration of $\gamma = \frac{1}{2}$ and $\beta = \frac{1}{6}$, which results in a linear variation of the accelerations.

By assuming the values of γ and β, and iterating Eq. 3.48, the structural responses can be calculated for the next time-steps given the current structural responses which is an appropriate method for the computer. The code for the Newmark-β analysis is presented in Appendix A.

3.7 Modal Analysis

Modal analysis is a fundamental technique used in structural engineering to investigate the dynamic behavior of structures. By examining the natural modes of vibration and corresponding frequencies, modal analysis provides valuable insights into the structural response under different loading conditions. This section focuses on the application of modal analysis specifically to structures equipped with TMDs. The modal analysis enables us to identify the critical modes that contribute significantly to the structural response and determine the optimal parameters for TMD placement and tuning. Through a detailed examination of the modal properties, including natural frequencies, mode shapes, and modal damping ratios, we can gain a comprehensive understanding of the dynamic characteristics of the structure. This knowledge is essential for designing effective TMD systems and achieving optimal vibration control and structural performance. In this section, we will explore various techniques and methodologies related to modal analysis, providing practical guidance for engineers and researchers in their analysis, design, and optimization of TMD-equipped structures.

Mathematically speaking, modal analysis transforms the equation of motion, which is a coupled set of equations that can be difficult to solve directly, into a set of uncoupled equations. This transformation simplifies the analysis process significantly. With this in mind, let us start with the equation of motion:

$$[M]\{\ddot{U}\} + [C]\{\dot{U}\} + [K]\{U\} = \{P(t)\}$$

By multiplying the displacement, velocity, and acceleration vectors with the stiffness, damping, and mass matrices, we obtain a coupled set of equations, which means that we cannot solve each equation independently.

To transform this coupled set of equations into an uncoupled form, we employ a technique similar to estimating a function using the Taylor series. This allows us to approximate the value of a complex function by summing simpler and deterministic terms. The more terms we consider, the more accurate the estimation becomes.

Eigen Vectors and Eigen Values

Returning to the equation of motion, in modal analysis, we identify unique displacement vectors known as **eigen vectors**. These eigen vectors can represent displacement vectors of the structure when combined through summation.

In mathematics, an eigen vector is a non-zero vector that remains unchanged in direction, but may be scaled by a factor known as the eigen value when multiplied by the transformation matrix. When a transformation matrix (such as a rotational matrix in a 2D plane) is multiplied by an eigen vector, the resulting vector remains in the same direction but may have a different magnitude. The factor that represents the effect of the transformation matrix on the eigen vector is called the eigen value.

The eigen vectors of a transformation matrix can be determined by:

$$[A]\{v\} = \lambda I\{v\} \tag{3.50}$$

or:

$$[A]\{v\} - \lambda I\{v\} = 0 \tag{3.51}$$

Therefore:

$$|A - \lambda I| = 0 \tag{3.52}$$

where:
$[A]$ is a transformer matrix
$\{v\}$ is the eigen vector for the matrix A
λ is the eigen value of matrix A
I is Identity matrix : $I_n = Diag(1, 1, ..., 1)$

By solving the equation mentioned, we can determine the eigen values of the transformation matrix. Substituting these eigen values into Eq. 3.51, we can calculate the corresponding eigen vectors. In most cases, the number of eigen vectors for a given transformation matrix is equal to the number of eigen values and is also equal to the dimension of the matrix. We can represent all the eigen vectors in a matrix $[V]$, where each column corresponds to a specific eigen vector.

Eigen vectors possess orthogonality, meaning that any other vector can be expressed as a linear combination of the eigen vectors. For instance, if v_1 and v_2 are the eigen vectors of the transformation matrix A, then any vector v can be written as the sum of these factored eigen vectors:

$$v = a \cdot v_1 + b \cdot v_2 \tag{3.53}$$

In structural analysis, instead of a transformation equation 3.51, we have a mathematical equation that the eigen values and eigen vectors must satisfy, which is the equation of motion. If we denote the eigen vectors of the structure as $[\phi]$, we can express any displacement vector as a sum of the factored eigen vectors. The factors are independent of time in a static analysis, while in a dynamic analysis, they become time-dependent.

$$\{u(t)\} = \sum_{i=1}^{N} \{\phi_i\} \cdot q_i(t) \tag{3.54}$$

In this equation,ϕ_i represents the displacement vector in the ith mode, similar to an eigen vector, and $q_i(t)$ is the time-dependent factor associated with ϕ_i. By obtaining the modal displacements, also known as mode shapes, at each time-step, we can calculate the values of $q_i(t)$ for all N modes. Multiplying these factors by their corresponding mode shape (ϕ_i) and summing them gives us the displacement vector of the structure at that specific time-step.

Thus, we have two unknowns in these calculations:

1. Mode shapes: These are the constant parts that remain the same across all time-steps.
2. The values of $q_i(t)$ for each time-step, representing the time-dependent factors.

Mode Shapes

The mode shapes of a structure play a crucial role in understanding its dynamic behavior. They represent the distinct patterns of motion exhibited by the structure in response to dynamic loads. Understanding mode shapes provides valuable insights into the dynamic characteristics of structures and aids in developing effective strategies for vibration control and structural optimization. In order to find the mode shape vectors, we begin by establishing the relationship between the mass and stiffness matrices. This relationship allows us to solve the eigenvalue problem and obtain the mode shapes and corresponding eigenvalues.

As discussed, the mode shapes are generally the eigen vectors of the structure.

In order to find the mode shape vectors we begin by writing the relationship between the mass and stiffness matrix is as follow:

$$\omega = \sqrt{\frac{k}{m}} \rightarrow k = \omega^2 \cdot m \tag{3.55}$$

$$\frac{k}{m} - \omega^2 = 0 \tag{3.56}$$

In a multi degree of freedom system, the k and m would change to the matrices. By multiply both sides to the displacement vector:

$$\frac{[K]}{[M]} \{\phi\} - \omega^2 I \{\phi\} = 0 \tag{3.57}$$

By comparing this equation to Eq. 3.51, it can be understood that the mode shapes of the structure, ϕ 's, are the eigen vectors of the $\frac{[K]}{[M]}$ matrix while the eigen values, ω^2, are the squares of the modal angular frequencies. As discussed, the eigen values, ω^2, can be calculated by solving following equation:

$$\left|[K] - \omega^2 [M]\right| = 0 \tag{3.58}$$

Having the modal frequencies, the mode shapes can be determined easily by solving Eq. 3.57 for each ω_i^2 value. As result, for each mode i, we would have a ω_i^2, and a the corresponding displacement vector, ϕ_i. Generally, the calculated mode shapes shall be normalized which means that the vector will be divided by the maximum of its absolute values. The mode shape vectors are orthogonal to each other, meaning they represent independent patterns of motion.

Orthogonality of Modes

The natural modes of a structure which are corresponding to different natural frequencies have orthogonality property which means that:

$$\{\phi_i\}^T [M] \{\phi_j\} = 0 \tag{3.59}$$

$$\{\phi_i\}^T [K] \{\phi_j\} = 0 \tag{3.60}$$

If $i \neq j$ and $\omega_i \neq \omega_j$.

The orthogonality of modes makes the following matrices diagonal:

$$[M_\phi] = [\Phi]^T [M] [\Phi] = 0 \tag{3.61}$$

$$[K_\phi] = [\Phi]^T [K] [\Phi] = 0 \tag{3.62}$$

The matrix $[\Phi]$ is the mode shape matrix in which each row is related to a DOF in the structure and each column is related to a mode shape. The matrices K_ϕ and M_ϕ are the modal stiffness and mass matrices where each ith diagonal element is related to the modal stiffness and mass for the ith mode. The modal mass, damping, and stiffness matrices can be also calculated for the ith mode as follow:

$$m_i = \{\phi_i\}^T [M] \{\phi_i\} \tag{3.63}$$

$$c_i = \{\phi_i\}^T [C] \{\phi_i\} \tag{3.64}$$

$$k_i = \{\phi_i\}^T [K] \{\phi_i\} \tag{3.65}$$

The stiffness and the mass of the ith mode are scalar values and like SDOF systems, they are related as follow:

$$k_i = \omega_i^2 m_i \tag{3.66}$$

Response History Analysis

The findings so far have demonstrated that the motion equation for a system with multiple degrees of freedom can be separated using eigen vectors, also known as the structure's mode shapes. Let's formulate the equation for a system with multiple degrees of freedom(MDOFs) undergoing earthquake-induced ground motions:

$$[M]\{\ddot{U}\} + [C]\{\dot{U}\} + [K]\{U\} = \{P_{eff}(t)\} \tag{3.67}$$

In this equation, $P_{eff}(t)$ is the loads due to external excitation (e.g. earthquake) on the degrees of freedom that can be affected by such external excitation. For earthquake excitation:

$$\{P_{eff}(t)\} = -[M]\,I\ddot{u}_g(t) \tag{3.68}$$

I is Identity matrix. It should be noted that in case that not all the DOFs of the structure are in the direction of the ground motion, or when the earthquake excitations is not identical for all the supports the general form for the external excitation should be used:

$$\{P_{eff}(t)\} = -[M]\{\iota\}\,\ddot{u}_g \tag{3.69}$$

Therefore the equation of the motion for a earthquake excited system can be written as:

$$[M]\{\ddot{U}\} + [C]\{\dot{U}\} + [K]\{U\} = -[M]\{\iota\}\,\ddot{u}_g(t) \tag{3.70}$$

The vector ι is the influence vector that makes the dynamic load to be applied only on the DOFs that can be affected by that. An explanation for better understand is that, if we divide the displacement vector of each mass into two components:

$$u_j(t) = u_j^s + u_j^t \tag{3.71}$$

Whereu_j^s is the quasi-static component which is the displacement due to static application of external excitation and u_j^t is the dynamic displacement component. The quasi-static component can be expressed as:

$$u_j^s = \iota\ddot{u}_g \tag{3.72}$$

In this case, the influence vector ι signifies the displacements of the masses that occur due to the static application of a unit ground displacement. In order to better understand the influence vector ι consider a frame with four DOFs as it is shown in Fig. 3.4. In this frame each mass has one translational and one rotational DOF. By applying the unit support displacement, the displacements in DOFs represents the ι vector. Therefore, the influence vector will be:

$$\iota = \begin{Bmatrix} 1 \\ 0 \\ 1 \\ 0 \end{Bmatrix}$$

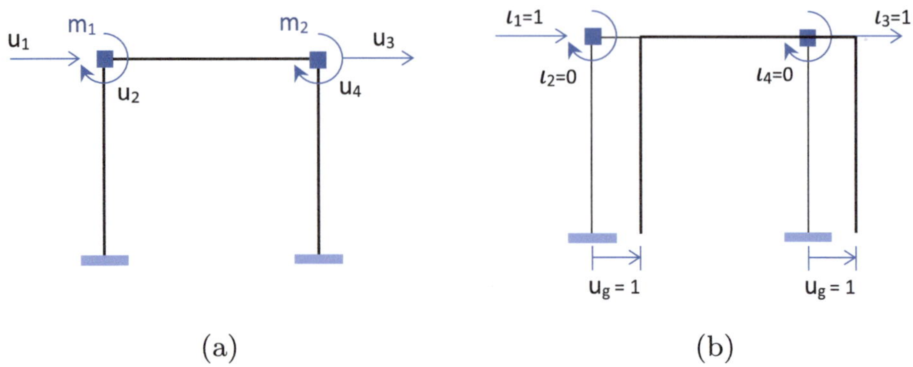

Fig. 3.4 **a** Multi degree of freedom system **b** Determination of ι values for each DOF based on the unit support displacement in earthquake excitation direction

Modal Equations

The preceding sections introduced the concept of a structure with MDOFS and its displacement vector. An essential cornerstone of modal analysis is understanding that this displacement vector can be represented as a cumulative assembly, or superposition, of modal contributions. This superposition is mathematically expressed as:

$$\{u(t)\} = \sum_{i=1}^{N} \sum \{\phi_i\} \cdot q_i(t) \tag{3.73}$$

This equation serves as a linchpin of modal analysis. The comprehension of this principle is pivotal, as subsequent derivations and methodologies essentially hinge on the substitution of this term into the motion equation. The result of such a substitution transforms a complex, intertwined set of differential equations into a more manageable, discrete set of equations.

From a mathematical perspective, the equation calculates a time-dependent function, rendered as a factored sum of harmonic functions. A noteworthy detail to consider here is that in this equation, $q_i(t)$ symbolizes a harmonic function that is related to the ith mode. This understanding and interpretation of the equation further elucidate the underlying principles and applications of modal analysis within the sphere of structural seismic control systems.

Substituting Eq. 3.73 in equation of the motion for a earthquake excited system:

$$\sum_{i=1}^{N} \sum [M]\{\phi_i\}\ddot{q}_i(t) + \sum_{i=1}^{N} \sum [C]\{\phi_i\}\dot{q}_i(t) + \sum_{i=1}^{N} \sum [K]\{\phi_i\}q_i(t) = \left\{P_{eff}(t)\right\} \tag{3.74}$$

By multiplying the Eq. 3.74 by $\left\{\phi_n^T\right\}$:

$$\sum_{i=1}^{N}\sum\left\{\phi_n^T\right\}[M]\{\phi_i\}\ddot{q}_i(t)+\sum_{i=1}^{N}\sum\left\{\phi_n^T\right\}[C]\{\phi_i\}\dot{q}_i(t)+\sum_{i=1}^{N}\sum\left\{\phi_n^T\right\}[K]\{\phi_i\}q_i(t)=\left\{\phi_n^T\right\}\{P_{eff}(t)\}$$

Considering the orthotropic property of the modes, the terms inside the sums will have zero value when $i \neq n$. Therefore, the equation will be simplified as follow:

$$\phi_n^T[M]\{\phi_n\}\ddot{q}_n(t)+\left\{\phi_n^T\right\}[C]\{\phi_n\}\dot{q}_n(t)+\left\{\phi_n^T\right\}[K]\{\phi_n\}q_n(t)=\left\{\phi_n^T\right\}\{P_{eff}(t)\} \tag{3.75}$$

The Eq. 3.75 can be derived considering the previously defined modal mass, damping, and stiffness as they defined before:

$$m_n\ddot{q}_n(t)+c_n\dot{q}(t)+k_nq(t)=\left\{\phi_n^T\right\}\{P_{eff}(t)\} \tag{3.76}$$

Substituting the effective external forces in case of earthquake excitation using Eq. 3.69:

$$m_n\ddot{q}_n(t)+c_n\dot{q}(t)+k_nq(t)=-\left\{\phi_n^T\right\}[M]\{\iota\}\ddot{u}_g(t) \tag{3.77}$$

The earthquake loads can also be expressed as summation of the modal inertia forces:

$$-[M]\{\iota\}\ddot{u}_g=-\sum_{i=1}^{N}\sum\Gamma_i[M]\{\phi_i\}\ddot{u}_g(t) \tag{3.78}$$

$$\Gamma_i=\frac{L_i}{M_i} \tag{3.79}$$

$$L_i=\{\phi_i\}^T[M]\{\iota\} \tag{3.80}$$

$$M_i=\{\phi_i\}^T[M]\{\phi_i\} \tag{3.81}$$

By substituting the inertia forces in the Eq. 3.77 and considering the orthotropic properties of modes:

$$m_n\ddot{q}_n(t)+c_n\dot{q}(t)+k_nq(t)=-\Gamma_nm_n\ddot{u}_g(t) \tag{3.82}$$

Thus far, we have converted the original motion equation into an uncoupled set of equations. This transformation was accomplished by approximating the displacement of the structure as a weighted summation of the modal displacement. Using the orthotropic feature of the modes, we could write the equation of the motion for the nth mode which is similar to a SDOF system. The equation of motion for an SDOF system is more conventionally written in the following form:

$$\ddot{q}_n(t)+2\zeta_n\omega_n\dot{q}_n(t)+\omega_n^2q_n(t)=-\Gamma_n\ddot{u}_g(t) \tag{3.83}$$

Solving the Eq. 3.83 will results in having $q_n(t)$ for the external excitation $\ddot{u}_g(t)$. The displacement of the masses in the nth mode can be obtained easily:

$$\{u_n(t)\} = \{\phi_n\}\, q_n(t) \tag{3.84}$$

Having the modal displacement helps us to calculate other parameters:

Total Response

having the modal responses for each mode, the total displacement vector of the masses can be obtained by summing the modal displacements:

$$\{u(t)\} = \sum_{i=1}^{N} \sum \{\phi_i\} \cdot q_i(t) \tag{3.85}$$

Having the displacements, the internal forces can be calculated by two procedures.

1. In the first variant, the internal forces are determined by using the structural stiffness matrix and the calculated displacements at DOFs. In case that the number of DOFs are reduced to the dynamic DOFs using technics like static-condensation method, the displacements at static DOFs can be determined from the calculated displacements for dynamic DOFs.
2. The alternative is the equivalent static forces. In this procedure, the external force $f_n(t)$ that produces the displacement $u_n(t)$ at any time instance t is applied to the structure in each mode and the internal forces will be calculated in each time instance by performing static analysis of the structure under f_n forces.

In both instances, the internal forces are calculated for each mode independently. The components of internal forces for any given element are subsequently determined by aggregating all the corresponding modal internal forces.

$$r(t) = \sum_{i=1}^{N} \sum r_n(t) \tag{3.86}$$

In case of earthquake excitation, it is convenient to use the second procedure which matches more to the code based design. In this procedure, we reformulate the Eq. 3.83 by defining $D_n(t)$ as:

$$D_n(t) = \frac{q_n(t)}{\Gamma_n} \tag{3.87}$$

$$\ddot{D}_n(t) + 2\zeta_n\omega_n \dot{D}_n(t) + \omega_n^2 D_n(t) = -\ddot{u}_g(t) \tag{3.88}$$

Where $[K]$ is the stiffness matrix of the structure. The pseudo acceleration is defined as follow:

$$A_n(t) = \omega^2 D_n(t) \tag{3.89}$$

The modal component of the force distribution vector is defined by s_n:

$$s_n = \Gamma_n [M] \phi_n \tag{3.90}$$

Using $A_n(t)$, the internal forces in each time instance for nth mode can be calculated in each time instance t :

$$r_n(t) = r_n^{st} A_n(t) \tag{3.91}$$

where r_n^{st} is the internal force in the nth mode under s_n . Having the internal force for each mode, the value of internal force for each internal force is the sum of the corresponding values using Eq. 3.86.

The equivalent static force procedure can be summarized as follow:

1. Define the ground motion acceleration $\ddot{u}_g(t)$ for each time step with Δt intervals.
2. Determine the structural matrices $[M]$ and $[K]$.
3. Determine the modal damping ratios ζ_i .
4. Calculation of the natural frequencies ω_i and natural modes vibrations ϕ_i.
5. Calculate the modal components of the force distribution vector s_i for each mode.
6. For each mode $i \in [1, N]$ at each time step t:

   - Calculate internal forces R_i^{st} by applying s_n as external force using static analysis.
   - Calculate the pseudo acceleration $A_i(t)$ for each mode using Eqs. 3.88 and 3.89.
   - Calculate the partial internal force $r_i(t) = r_i^{st} A_i(t)$

7. Sum the partial internal forces for all the considered modes to calculate internal force component $r(t)$.

It should be noted that the presented formulation for modal analysis is suitable for calssically damped linear systems.

3.8 Check the Structure for Resonance Issues

Resonance is a critical phenomenon in the field of structural engineering that arises when a structure's natural frequency aligns with the frequency of an external dynamic load or environmental vibration. This alignment can result in significant amplification of the structure's response, potentially leading to catastrophic failure. Consequently, it is essential to conduct a thorough evaluation of resonance effects during the design and analysis phase of structural engineering.

Identification of Natural Frequencies: The first step in checking resonance effects involves determining the structure's natural frequencies. These are the frequencies at which the structure tends to response when subjected to external excitations. The identification of these frequencies can be achieved through theoretical calculations or experimental modal analysis methods.

Comparison with Load Frequencies: After identifying the structure's natural frequencies, the next step involves comparing these frequencies with the frequencies of potential dynamic loads or environmental vibrations the structure is likely to encounter. This could include machinery-induced vibrations, seismic loads, wind-induced vibrations, traffic-induced vibrations, and more.

Resonance Evaluation: If the structure's natural frequencies related to the first modes align with the frequencies of potential external influences, a resonance condition is possible. Here, it is vital to evaluate the potential impact of resonance on the structure's overall performance, taking into account the amplification of vibrations and the structure's damping characteristics. Advanced dynamic analysis techniques, such as response spectrum analysis or time-history analysis, can be used to quantify the potential effects of resonance.

Mitigation Measures: If the evaluation reveals a significant risk of damaging resonance, various mitigation measures can be considered. These could include modifying the structure to change its natural frequencies, introducing damping mechanisms to reduce vibration amplification, or altering the frequencies of the dynamic loads, if possible.

Under certain circumstances, due to the complexity of dynamic loads and/or the intricate characteristics of the structure, discerning how to reduce excessive deformations caused by external dynamic loads can be challenging. When considering a structure as a system under dynamic loads, four key parameters are in play: mass, stiffness, damping, and geometry. A pertinent question arises: which of these factors has the most significant influence on the structure's behavior under a specific dynamic load? To answer this, it is essential to comprehend the critical role each structural parameter plays compared to the others. Analyzing a SDOF system under a harmonic load P can provide valuable insights.

The harmonic load can be represented as:

$$P = P_0 \sin(\omega t) \tag{3.92}$$

The peak response of the SDOF system largely hinges on the ratio of the load frequency to the natural frequency of the SDOF system ($\frac{\omega}{\omega_n}$). The peak responses can be categorized into three primary scenarios:

1. Slow-Acting Load: When $\frac{\omega}{\omega_n} \ll 1$, this signifies that the period of the load, T is significantly longer than the structure's natural period. In other words, the load is acting very slowly on the structure:

$$u_0 = \frac{P_0}{K} \tag{3.93}$$

Here, P_0 denotes the peak of the harmonic load, and K is the stiffness of the SDOF system. This suggests that as the load frequency lowers relative to the structure's frequency, the peak response of the structure approximates the static displacement under the peak magnitude of the load.

2. Rapidly Changing Load: When $\frac{\omega}{\omega_n} \gg 1$, this indicates that the period of the load T is much shorter than the structure's natural period. In other words, the load is changing its direction very rapidly:

$$u_0 = \frac{P_0}{M\omega^2} \tag{3.94}$$

This demonstrates that the system's mass has the most substantial impact on the peak responses under the dynamic load.

3. Resonance Conditions: When $\frac{\omega}{\omega_n} \simeq 1$, this implies that the load's period T is close to the structure's natural period, which is a condition where resonance can occur:

$$u_0 = \frac{P_0}{2\zeta} = \frac{P_0}{C\omega_n} \tag{3.95}$$

In resonance situations, damping plays a crucial role and regulates the magnitude of the peak structural responses.

Introducing a response factor R, which is the ratio of the peak dynamic response to the static response, allows the representation of all three response ratios of the SDOF system—including peak ratios of deformation, velocity, and acceleration responses—in a single graph (Fig. 3.5). This graph provides a comprehensive overview to understand which structural parameter predominantly impacts total peak responses under a specific loading condition. For practical utilization of the graph, the point of the damping ratio graph is first determined

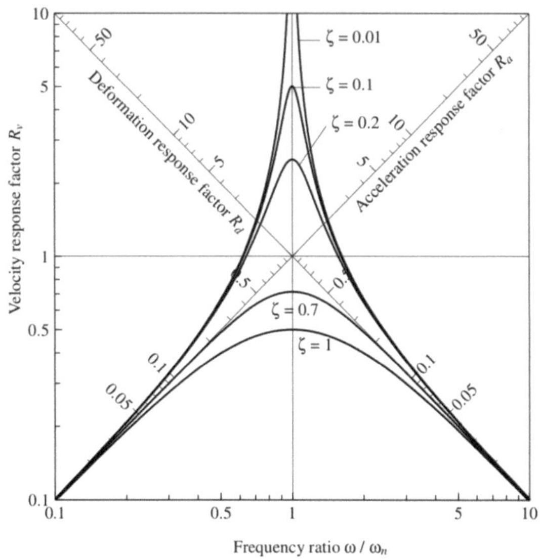

Fig. 3.5 Four-way logarithmic plot of deformation, velocity, and acceleration response factors for a damped system excited by harmonic force [5]

based on the load frequency ratio $\frac{\omega}{\omega_n}$. Subsequently, each peak response ratio can be read based on the coordinates of the graph's point on the corresponding axis.

In real-world projects, engineers often have to deal with structures that possess multiple degrees of freedom and are subject to non-harmonic dynamic loads, such as winds and earthquakes. These projects typically present two main challenges:

1. Investigating and evaluating structural behavior under external excitations.
2. Identifying an optimal and feasible solution to improve structural behavior.

The first challenge can generally be addressed by employing computer modeling and conducting dynamic analysis. For multi-degree-of-freedom systems, numerical solutions are commonly used to determine structural responses. One widely accepted method is the Newmark's-β method, which is further explained in this chapter.

The second task can be substantially more demanding. A traditional solution involves acquiring more comprehensive information about the dynamic characteristics of the structure. Executing a modal analysis offers a thorough overview of the natural modes and their frequencies. As discussed earlier, each mode, similar to a SDOF system, has a unique frequency, stiffness, mass, and damping value. It's customary to arrange the modes based on their frequency values, with the first mode having a lower frequency value compared to the others. Given that frequency is directly related to stiffness, a lower natural frequency implies reduced stiffness. Consequently, excessive deformations in structures are typically more associated with their responses in the first modes.

External loads also possess a frequency content, meaning that if we transform the load from the time-domain to the frequency-domain, the majority of external excitations may occur at one or more specific frequencies, with lower magnitudes at other frequencies. These dominant frequencies are crucial. When these frequencies align with a natural mode of the structure, it's said that the external load activates those modes. If a prominent frequency of the external load falls between two natural frequencies of the structure, the lower frequency can conservatively be considered the activated mode. This comparison grants the engineer additional insight into potential resonance frequencies.

In instances where these frequencies approach the first natural modes, the resonance effects can be more significant due to the structure's reduced stiffness in these modes, which could result in excessive deformations. In cases where the load and structure share the same frequencies in the higher modes, the magnitude of the load frequency content becomes important for considering that frequency and the corresponding mode in the list of critical modes to be controlled.

If resonance with one or more structural modes is detected through modal analysis, and the dynamic analysis reveals excessive structural responses, it then becomes essential to determine how to address these resonance effects. Generally, there are two primary strategies:

1. Frequency Shifting
2. Damping Improvement

'Frequency shifting' involves adjusting the frequency of either the external excitation or the structure to avoid resonance in the first modes. Shifting the external load's frequency might be feasible when the source of the external dynamic load is a controllable factor, such as equipment—a situation often encountered in industrial structures. Various techniques can be used to shift or dampen external excitations, such as employing isolators beneath the equipment's supports or constructing a secondary structure between the equipment and the primary structure. When dealing with wind or earthquake forces, frequency shifting can occur on the structure's side. This necessitates altering the mass and/or stiffness of the structure, a technique sometimes applied to bridges under wind and traffic loads. However, frequency shifting can be prohibitively expensive in some cases, making alternative solutions like damping improvement a more viable choice.

As demonstrated, damping plays a critical role in resonance conditions. Therefore, increasing the structural damping of the relevant mode or modes using certain types of dampers can considerably enhance the structural responses under resonance effects. Tuned Mass Dampers are one of the most commonly used dampers for mitigating excessive movements in cases of resonance.

In conclusion, checking for resonance effects in a structure is a vital aspect of structural design and analysis. This process aids in ensuring the structure's safety, durability, and overall performance in response to dynamic loads and environmental vibrations. Past experiences

Fig. 3.6 Collapse of Tacoma Narrow Bridge (1940)

highlight how even large structures can be vulnerable under resonance effects, as exemplified in (Fig. 3.6).

3.9 Ground Motion Selection

The selection of appropriate ground motions is a critical part of seismic analysis and design in structural engineering. Ground motion records are used as inputs to dynamic analyses of structures to assess their response to seismic loads. The selection process should ensure that the chosen ground motions are representative of the seismic environment to which the structure is exposed.

The American Society of Civil Engineers (ASCE) provides guidance on this subject through its seismic design standards. According to ASCE, there are several key parameters to consider when selecting ground motions:

1. **Magnitude and Distance:** The magnitude of an earthquake and the distance from the site to the source of the earthquake significantly influence the characteristics of ground motion. The selected ground motions should match the maximum considered earthquake ground motions used for the seismic hazard analysis.

2. **Site Conditions:** Ground motion is considerably influenced by the local geological and soil conditions. ASCE recommends considering the site's soil profile and the effect of local geology on the seismic wave's propagation.
3. **Spectral Characteristics:** The selected motions should be spectrally matched to the design spectrum. The spectrum of ground motion is a graphical representation of the maximum ground motion (displacement, velocity, or acceleration) as a function of the structure's period of vibration. Spectrally matching the ground motion ensures that the chosen records adequately represent the range of potential responses for structures of different dynamic characteristics.
4. **Fault Mechanism:** The nature of the fault (strike-slip, reverse, normal, etc.) that generates an earthquake influences the type of ground motion produced. Selected motions should be consistent with the fault mechanisms considered in the seismic hazard analysis.
5. **Directionality:** Earthquakes produce ground motions in three dimensions. While the vertical component of ground motion is often less severe than the horizontal components, all three dimensions should be considered when selecting ground motions.
6. **Number of Ground Motions:** ASCE recommends using a minimum number of ground motion pairs (usually at least seven) for a proper statistical representation of the structural response.
7. **Scaling of Ground Motions:** Ground motion records should be appropriately scaled to represent the intensity of shaking defined by the design spectrum. However, the scale factor should not excessively distort the original ground motion's significant features.

As mentioned, ASCE has specific recommendations regarding the number of earthquake records to be considered for a meaningful seismic analysis. According to ASCE 7, a minimum of seven ground motion pairs should be used for a structure's seismic design. Each ground motion pair typically consists of two orthogonal horizontal components of motion. Using this many ground motion pairs allows for a sufficient statistical representation of potential seismic responses. After selecting the earthquake records, they should be scaled based on a target response. Adjusting earthquake records to match a target response spectrum is a significant step in the seismic analysis of structures. This adjustment serves a number of important purposes:

- Relevance to Location: The ground motion recorded in an earthquake event is highly location-specific, reflecting the unique seismic characteristics and geological conditions of that area. Therefore, an earthquake record obtained from a different region may not accurately represent the seismic hazards at the site of the structure being analyzed. By adjusting the earthquake records to match a target response spectrum, which has been defined according to the local geological conditions and seismic activity, the analysis becomes much more relevant and accurate.

- Ensuring Uniformity: In seismic analyses, it's not unusual to use multiple ground motion records to account for variability in potential earthquake events. Adjusting these records to match the target response spectrum helps to ensure that they all reflect the same level of seismic intensity, making for a more consistent and fair comparison of the structure's response to different earthquake events.
- Consideration of Critical Frequencies: The target response spectrum is usually derived from the site's seismic hazard analysis, and it essentially captures the ground motion characteristics at different frequencies that are considered to be critical for the location. Adjusting the earthquake records to this spectrum helps ensure that the simulated ground motions are in line with these critical frequencies, leading to a more accurate evaluation of the structure's seismic response.
- Compliance with Codes and Standards: Many building codes and standards, like ASCE, require that the earthquake records used in seismic analyses match a certain target response spectrum that is specified by the code or standard. Hence, adjusting the records to these spectra ensures compliance with the relevant guidelines.

By carrying out these adjustments, engineers can significantly enhance the accuracy and reliability of their seismic analyses, resulting in designs that are better equipped to withstand potential earthquake events. In the following section the process of matching the earthquake records is presented.

3.10 Scaling and Matching Earthquake Records

The process of matching earthquake records is a crucial aspect in the realm of structural seismic analysis. The overarching aim is to adjust the earthquake accelerograms, which are recordings of ground motion during an earthquake event, so they align with a specific target response spectrum. This target response spectrum is typically defined in accordance with the local seismic hazard characteristics and the building code requirements of the site where the structure is located.

The need to perform this adjustment arises due to several important considerations:

1. **Geographical Relevance:** Earthquake records capture ground motion as it occurred in a specific location during a specific event. The geological conditions and seismicity of this location may greatly differ from those of the site where the structure being analyzed is situated. By adjusting the earthquake records to match a target response spectrum representative of the site-specific conditions, the seismic analysis becomes significantly more accurate and reliable.
2. **Uniformity in Analysis:** In any seismic analysis, it is commonplace to utilize multiple ground motion records. This practice acknowledges the inherent variability in potential seismic events. However, to make a fair comparison of the structure's response to

different earthquake records, these records need to be adjusted to reflect the same seismic intensity level. This is achieved by matching them to the target response spectrum.

3. **Highlighting Critical Frequencies:** The target response spectrum is usually informed by a seismic hazard analysis of the site, which identifies the critical frequencies that need to be considered for that specific location. By adjusting earthquake records to this spectrum, the ground motions used in the analysis are in line with these critical frequencies. This, in turn, leads to a more precise and robust evaluation of the structure's seismic response.
4. **Compliance with Building Codes and Standards:** Many building codes and standards, such as those defined by the ASCE, necessitate that the earthquake records used in seismic analyses match a predefined target response spectrum. Therefore, adjusting the records to these spectra ensures adherence to the relevant regulatory guidelines.

Several methods are available for matching earthquake records to a specific response spectrum. One straightforward technique is the iterative method, which uses an assumed scale factor for each seismic record. This scale factor is then fine-tuned by comparing the resulting response spectra to the target spectrum in an iterative process. While this method can be readily implemented, using a software application that simplifies data manipulation can be beneficial. Microsoft Excel is one such example.

The procedure of matching earthquake records initiates by deriving the spectrum for each individual earthquake acceleration record from the selected group. An average spectrum is then compiled and adjusted to align with the predetermined target response spectrum, which is based on the specific design code and site specifications. The alignment or matching process is accomplished by tuning the scale factors for each individual spectrum.

Here is a detailed, step-by-step guide to this process:

1. Define the Target Response Spectrum:Establish the target response spectrum based on applicable building codes.
2. Select Earthquake Acceleration Records: Choose the earthquake acceleration records from available databases like the Pacific Earthquake Engineering Research Center (PEER) Ground Motion Database.
3. Generate the Response Spectrum for Each Accelerogram: For each earthquake record, create a corresponding response spectrum. This spectrum plots the maximum response of a theoretical Single Degree of Freedom (SDOF) system to an input vibration as a function of natural frequencies. The responses can be in various forms, including displacement, velocity, or acceleration. The SDOF system's damping used to form the response spectrum should be determined based on the target response spectrum (typically 5%). When using Excel, dedicate one sheet to each response spectrum data for an earthquake.

4. Since records are often provided in two perpendicular directions, the resulting horizontal spectrum is derived from the Square Root of the Sum of Squares (SRSS) of the spectra for the two horizontal components within each set of earthquake records
5. For each individual horizontal spectrum, we assume a scale coefficient that minimizes the difference between the corresponding target spectrum amplitudes. Next, the amplitude of each of the resultant horizontal spectra is scaled by multiplying it with the corresponding scale coefficient.
6. Calculate the Average Response Spectrum: Determine the average of all the response spectra at each period.
7. Adjust the Scale Factor: Modify the scale factor for each earthquake record so that:

    - The spectra for all records be near the target response spectrum
    - The spectra for all records do not fall below the 1.3 times of the target spectrum over the period range from $0.2T_1$ to $1.5T_1$, where T_1 is the fundamental period of the structure.

8. Iterate Until Criteria are Met: Repeat the previous steps until the specified criteria are satisfied. This iterative process ensures that the response spectra of the selected and scaled seismic records adequately match the target response spectrum, thereby providing a more accurate and code-compliant assessment of the structure's seismic response.

In cases where this ratio exceeds 1.3, scale coefficients are determined for all selected sets of earthquake acceleration records, ensuring an optimized match between the recorded and target earthquake spectrums. A visual example of the matched earthquake records can be found in the following Fig. 3.7.

> Visualizing the iterative process can greatly facilitate earthquake record matching. Constructing a graph that includes the response spectra, the target response spectrum, and the factored response spectrum (which defines the permissible range for the average spectrum) can enhance the iterative process. This graphical representation provides a clear visual comparison, making it easier to determine whether the current iteration satisfactorily matches the target response spectrum or if further adjustments are needed.

Some commercially available software packages can automate the process of scaling and matching. For instance, SeismoMatch is a tool designed to match the response spectra of chosen earthquake records to a predetermined target spectrum. The following steps show the process utilizing SeismoMatch:

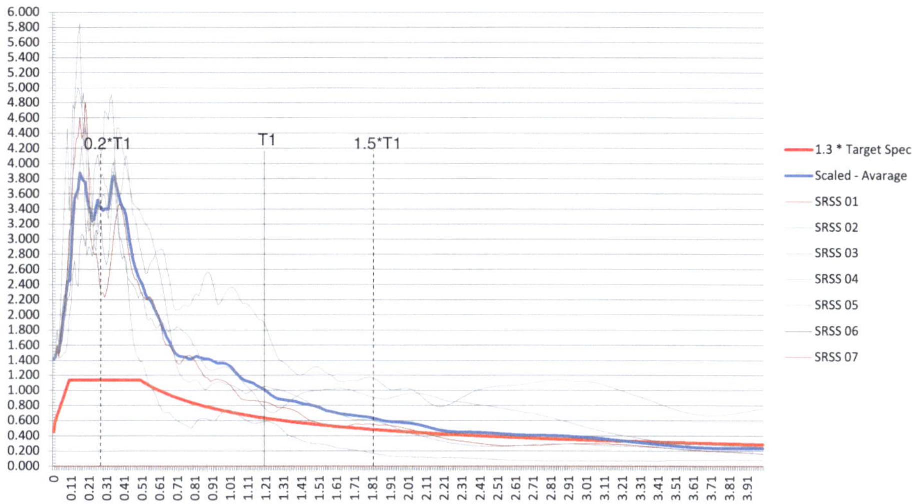

Fig. 3.7 Scaling and matching response spectrums to match a target response spectrum

1. Uploading Earthquake Records.
2. Inputting the Target Spectrum: Once your accelerogram files are uploaded, proceed to define your target spectrum. This can be done by navigating to the "Spectrum" menu and choosing the "Target Spectrum" option. The target spectrum information can either be loaded from a .txt or Excel file, or manually defined via the user interface.
3. Choosing the Matching Algorithm: SeismoMatch provides a range of spectral matching algorithms from which you can choose the most suitable one.
4. Running the Matching Process.
5. Analyzing the Results: Post completion of the matching process, SeismoMatch presents the adjusted and original accelerograms alongside the original and target response spectra in a plotted format for comparison. Examine these plots closely to ensure the adjusted accelerogram's response spectrum is satisfactorily matching the target spectrum.
6. Apply the scale factors to the acceleration records of the earthquakes to obtain the matched records.

It is important to note that in a seismic assessment study, the target spectrum may not necessarily align with the code-defined spectrum for the structure's location but may instead be tied to a user-specified hazard level.

3.11 Fourier Transformation: Bridging Time and Frequency Domains

In behavior investigation of a structure under earthquake excitations, we generally use dynamic analysis to evaluate the structural responses using direct methods in time-domain. At the end, both the external excitations and the structural responses are some raw data in time-domain. Transforming these data from time-domain to frequency-domain gives more inside to the frequency content of the data and helps us with the interpretations of the analysis results and making conclusions.

One of the effective methods for transforming a time-series into frequency domain is the Fourier transformation.

The Fourier Transform, named after the French mathematician and physicist Jean-Baptiste Joseph Fourier, is a powerful mathematical tool used extensively in numerous fields including seismic analysis. Its fundamental role is to convert a function from the time domain, where it represents a signal as a sequence of values over time, into the frequency domain, where the same signal is instead represented as a sum of sinusoidal components of varying frequencies.

In the context of seismic analysis, the time domain representation of a seismic event is an earthquake accelerogram, which is essentially a record of ground acceleration as a function of time. The Fourier Transform of an accelerogram generates a frequency spectrum, revealing the different frequencies present in the seismic event and their respective amplitudes.

The Fourier Transform's inverse function can convert back from the frequency domain to the time domain, making it a versatile tool for seismic analysis. It allows for the examination of seismic events from both the time perspective (capturing when events occur) and the frequency perspective (identifying which frequencies are most dominant). This dual capability is essential for a comprehensive understanding of the event's impact on structures.

The Fourier transformation of a function $f(t)$ in time domain is the function $F(\omega)$ in the frequency domain, where:

$$F(\omega) = \int_{-\infty}^{+\infty} F(t)e^{-i\omega x}\, dx \tag{3.96}$$

Recall that:

$$i = \sqrt{-1} \tag{3.97}$$

and:

$$e^{i\theta} = \cos\theta + i\sin\theta \tag{3.98}$$

The integral runs from negative to positive infinity, capturing the entirety of the function f(t). The result, $F(\omega)$, is a complex-valued function of frequency, with both magnitude and phase. The magnitude represents the amount of the frequency ω present in the original function, while the phase represents the shift in time of the sinusoid in that frequency. To understand this formula, imagine an arbitrary function where a sinusoid wave with a particular frequency comes and collides the function in a some points. This equation sums all those points and consider that as the value of the transformed function at that frequency for that arbitrary function. Now, by altering this frequency we can achieve the transformed function for all frequencies.

$$F(\omega) = \int_{-\infty}^{+\infty} F(t)e^{-i\omega x}\, dx \tag{3.99}$$

Discrete Fourier Transformation

In many engineering problems, we are not dealing with a continuous function of time, but rather a discrete function composed of a collection of data points over time. A common example in structural engineering is the roof acceleration response of a building in a particular direction, as collected through accelerometer sensors or obtained from a time-history analysis of a model. In such scenarios, we are presented with a set of time and acceleration values.

To analyze these kinds of problems, we use a special form of Fourier Transform called Discrete Fourier Transform (DFT). This tool allows us to transform a discrete function in the time domain into a discrete function in the frequency domain.

Let's consider a signal represented as a_n for $n = \{0,\ 1,\ \ldots,\ n-1\}$. The Discrete Fourier Transform of this signal can be calculated as follows:

$$A_k = \sum_{n=0}^{N-1} W_N^{Kn}\, a_n \tag{3.100}$$

where:

$$W_N = e^{-i\frac{2\pi}{N}} \tag{3.101}$$

The principle behind the Discrete Fourier Transform is essentially the same as that of the Fourier Transform, where two harmonic functions (Cosine and Sine) are utilized to capture the values of the main function and average them. The reason for the presence of these two periodic functions, with a phase difference of $\pi/2$, is to ensure that they cover for each other—when one of them reaches zero at a point, the other is at one of its extremes.

In practice, we generally use a fast and efficient algorithm, called the Fast Fourier Transform (FFT), to compute the Discrete Fourier Transform of a function. The FFT is derived by first considering the 2-point and 4-point DFTs and then generalizing it to 8-point, 16-point, and so on up to 2^r points. This approach significantly reduces the computational complexity, making FFT a crucial tool in digital signal processing and other related fields. Such functions are available in packages like MATLAB which is a versatile computational platform, provides an array of mathematical functions that facilitate the transition between time and frequency domains, further simplifying the analysis process. A code for computing FFT of an earthquake record in presented in Appendix A

3.12 Genetic Algorithm: A Bio-Inspired Optimization Technique for Engineering Solutions

Before diving into the application of Genetic Algorithm (GA), it is crucial to grasp the foundational concept of optimization, an essential process in numerous scientific and engineering fields. Optimization is a computational method used to find the best solutions or decision alternatives for a given problem within a set of constraints. These solutions are typically the ones that either maximize or minimize a particular objective or a set of objectives.

Broadly, optimization problems are divided into two main categories: single-objective optimization and multi-objective optimization. Single-objective optimization problems involve optimizing one objective function subject to certain constraints. This optimization finds an optimal solution that minimizes or maximizes this objective function. An example of a single-objective optimization problem might be finding the most cost-effective design of a structure given a set of constraints such as building codes, material availability, or construction techniques.

On the other hand, multi-objective optimization problems involve optimizing more than one objective function simultaneously, which more accurately represents real-world problems. In most practical applications, there are multiple conflicting objectives to be optimized. For instance, in the optimization of TMDs, there could be several objectives such as minimizing structural displacements, accelerations, and velocities simultaneously. Each of these objectives could potentially conflict with one another; minimizing displacements might result in higher accelerations, or reducing velocities might increase displacements. This multi-objective nature makes the problem more complex as there is no single solution that optimizes all objectives simultaneously. Instead, a set of optimal solutions, known as the Pareto optimal set, is sought, where no objective can be improved without degrading at least one other objective.

Understanding the type of optimization problem one is dealing with is essential as it informs the choice of the optimization algorithm. GA is particularly suited to multi-objective problems due to its ability to explore a large solution space and its inherent capacity to handle multiple objectives, making it an ideal choice for optimizing TMD parameters. This kind of optimization provides an opportunity to find a balanced solution that best fits the requirements of the design and operation of a TMD, ultimately improving the structural performance.

In the realm of complex problem-solving, researchers and developers often resort to advanced computational strategies. GA is one such powerful tool. As part of the broader family of evolutionary optimization methods, GAs draw inspiration from biological processes, particularly the principles of natural selection and genetics. Within the GA framework, potential solutions are depicted by *chromosomes*, each represented by a chain of numerical or character-based elements, known as *genes*. Each chromosome serves as a viable solution to the problem, embodying the values for all variables. Thus, the chromosome can be decoded to extract these specific values.

This coded representation, or chromosome, subsequently undergoes transformations facilitated by GA operators such as *crossover* and *mutation*. This iterative process modifies the structure of the chromosome, guiding the algorithm toward an optimal solution. Over time, these modifications gradually refine the solution set, optimizing the problem variables and facilitating the convergence to the most promising solution.

In this section, we will explore the conceptual basis of GA and highlight its utility in engineering contexts, especially in tuning the parameters of mass dampers.

Optimization lies at the heart of research and development, serving as a decisive element that differentiates adequate solutions from outstanding ones. With the intricacies of engineering problems requiring precise, robust, and efficient optimization techniques, GAs stand out due to their adept handling of complex problems involving numerous parameters and constraints.

In the context of GA, the problem at hand is dissected into objective functions, variables, and solutions. For instance, when optimizing TMDs, variables encompass parameters like mass, stiffness, and tuning frequency. The corresponding values for these variables form the solutions, while the objective functions can be described as the ratio of uncontrolled to controlled structural responses. These functions, termed as objective because we input solutions to them and their outputs give us the values for the objectives, can be various performance metrics like roof displacement, velocity, or acceleration for a given set of TMD parameters.

Genetic Algorithms commence with an array of random solutions, often referred to as a *population*. The *individuals* in this population, essentially different solutions, evolve over time towards better outcomes. Each candidate solution, or individual, is encapsulated within a *chromosome* comprising an encoded set of variables.

The evolution process iteratively applies genetic operations to the individuals, creating new members and selecting elite members to form successive generations. This new generation of solutions then feeds into the next iteration of the algorithm. Termination of the algorithm typically happens upon reaching a predetermined number of generations or a desired fitness level.

The *fitness function* in GA serves as a measure of the goodness of each solution, enabling the algorithm to compare and rank solutions, eliminate less desirable ones, and utilize the better solutions to evolve improved ones. Hence, understanding and applying such optimization techniques becomes vital for researchers, particularly in fields like structural engineering, where the optimization of design parameters can profoundly impact the safety and performance of structures. Tuned mass dampers provide a perfect use case for Genetic Algorithms, exemplifying how they can effectively tackle the optimization of multi-parameter systems.

In the following sections, we delve into the mechanisms of the GA, illustrating its functioning and utility in the realm of engineering optimization. Readers will learn about the algorithm's components, its workflow, and how it can be leveraged to optimize the parameters of tuned mass dampers effectively. The intention is to equip the reader with an understanding and appreciation of GA, fostering their ability to tackle complex optimization problems in their research work.

3.12.1 Variable Preparation and Selection

The first and critical step in tackling optimization problems is defining and preparing the variables. Although this task might appear straightforward, it's not always the case. Often, some preparatory work on the variables significantly enhances the efficiency of the optimization process. This is especially pertinent when employing GA, as it is advisable to ensure that both the variables and the fitness value(s) are dimensionless.

Consider, for instance, the optimization of TMDs, where the variables include the mass, stiffness, and damping of the TMDs. Each of these parameters has its own unit of measure (e.g., Kg for mass, N/m for stiffness, N.s/m for damping). One strategy for rendering these dimensionless is to normalize each by the corresponding values for the structure. Hence, the mass ratio of the TMD to the structure becomes one of the dimensionless optimization variables. Similar transformations can be applied to the fitness functions—for instance, considering the ratio of controlled to uncontrolled displacements as the objective function, which is dimensionless. Experience suggests that such initial preparations of the variables can considerably enhance the performance of the algorithm.

Another technique in variable preparation is the elimination of dependent variables. All variables considered in the optimization process should be independent, meaning if a variable can be derived from others, it should be omitted from the optimization. To put it another way, the optimization process should start with an initial set of independent variables. For complex problems, a mapping may be employed to correlate some independent variables, which itself might require preliminary mathematical work.

A third method of variable preparation involves scaling the variables. Here's why and how. In GA, we employ coding and decoding to convert the value of a variable into a series of numbers or letters, representing the chromosomes. The size of this chromosome depends on the coded value for each variable and each solution. Generally, the total size of the chromosome is constant and determines the range of the coded solution. In practice, the size of the chromosomes significantly impacts the performance and convergence speed of the algorithm by reducing the number of null or non-feasible solutions. Hence, it is advised that the chromosome size be kept as small as possible. Since the potential maximum and minimum of all variables control the size of the chromosome, having variables within similar ranges helps the chromosome size be used effectively. Therefore, variables are sometimes scaled to fit within a range comparable to others, and after the optimization process concludes, they are rescaled back to their actual values.

3.12.2 Encoding

In the Encoding section, we delve into the critical role of encoding in GAs, which is aimed at converting variable values into a chromosome format. A thorough understanding of encoding is essential for formulating an engineering problem within the GA framework, thus facilitating problem-solving.

In an optimization problem, we often contend with single or multiple variables, the optimal values of which we aim to identify. These variables can be multifarious in nature. For instance, when designing TMDs, the optimal solution includes determining the TMD's mass, damping, and tuning frequency. This combination of parameters will yield the TMD's optimal performance, given the structural properties.

In GA, we encode all these variables into a dimensionless, easily manipulatable string. This string is then subjected to modifications by GA operators. Ultimately, we obtain the optimal solution as a string, which can be decoded to yield the optimal values for each parameter.

Binary coding is one of the most frequently used methods in GA due to its compatibility with computer processing. In this approach, we usually determine a possible value range for each parameter and establish the length of the binary string that can accommodate these values. By combining the strings for each parameter, we construct an individual—a viable solution to the problem.

For instance, Fig. 3.8 provides a schematic representation of the binary coding process used to encode TMD parameters into a chromosome. As depicted, each variable is encoded into a binary code, and all three binary codes for the three variables are then amalgamated to form the chromosome.

> It's important to know that the length of the binary string, corresponding to each variable, must remain constant throughout the optimization process. This string length is determined based on the minimum and maximum possible values for each variable. Consequently, it's essential to define the feasible range for each variable beforehand.

3.12.3 Selection

In GA, the selection process serves as a crucial evolutionary mechanism, much like natural selection in the biological world. It operates on the principle of 'survival of the fittest', allowing for the perpetuation of only the most superior solutions or individuals from one generation to the next. This process aids in the progression and refinement of solutions, and as such, stands central to the operation of GA.

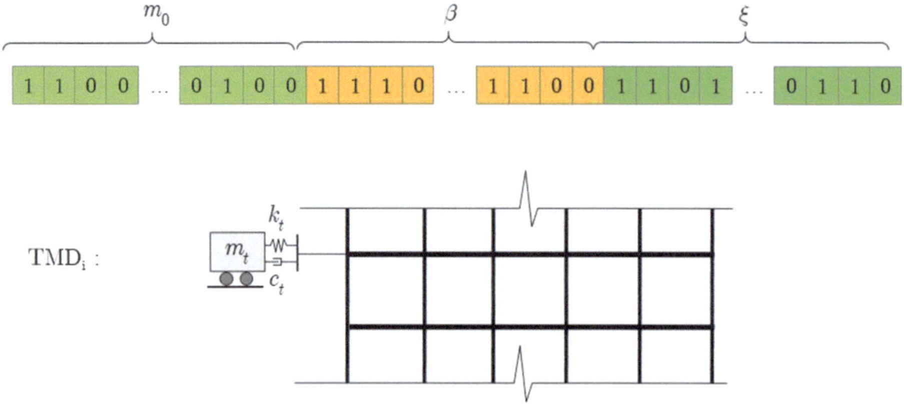

Fig. 3.8 Binary coding the TMD's parameters

Selection in GA is a two-step operation. First, the algorithm picks out solutions, or individuals, from the existing population based on their fitness levels. In this step, a bias towards superior solutions exists, allowing them a greater chance of being selected for reproduction. This bias, however, must be finely balanced; an excessive selection pressure can lead to premature convergence towards sub-optimal solutions, while too little can slow down convergence by rendering the search directionless.

In the second step, the chosen solutions are grouped into what is often referred to as the *mating pool*. This pool is then subjected to further genetic operations, such as crossover and mutation, to create offsprings for the subsequent generation. Through this process, the algorithm inches closer to finding the optimal solution with each successive generation.

One popular selection method used in GA is the roulette wheel selection. Illustrated in Fig. 3.9, this method likens the selection process to a linear search through a roulette wheel, wherein each slot represents an individual. These slots are weighted in accordance with the fitness value of their respective individuals, ensuring a higher chance of selection for fitter solutions [93]. This graphical representation underscores the basic principle of the selection process in GA—the survival and reproduction of the fittest. Regardless of the selection strategy employed, be it roulette wheel selection, tournament selection, or rank selection, the goal remains consistent: the pursuit of optimal solutions while maintaining a diverse population. Striking this balance enables the genetic algorithm to effectively navigate the solution space, ultimately enhancing its overall performance and convergence rate.

3.12.4 Genetic Operators

In GA, genetic operators are tools that allow the algorithm to navigate the solution space and drive the search towards optimality. These operators, reminiscent of natural genetic

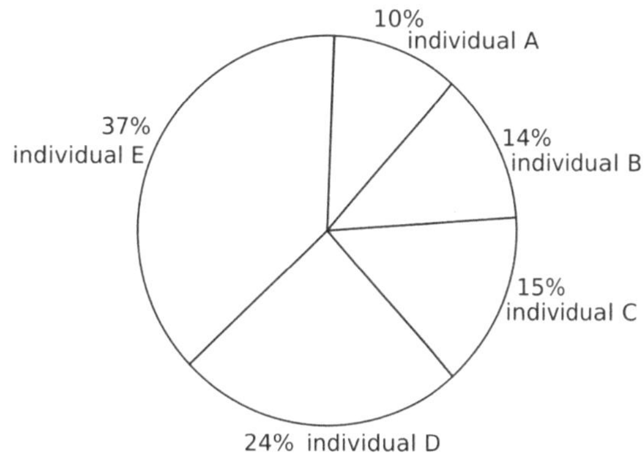

Fig. 3.9 Roulette wheel selection

mechanisms, perform crucial operations on the population of solutions, enhancing their fitness values over successive generations. The three primary genetic operators used in GA are crossover (or recombination), mutation, and selection.

1. Crossover (Recombination): This operator plays a significant role in exploring the search space by combining genetic material from two parent chromosomes to create new offspring. The process begins with the selection of two parent solutions from the population. A crossover point is then randomly chosen along the length of the chromosomes, and segments from the parents are swapped to produce new offsprings. The goal is to generate offspring solutions that possess superior traits from both parents, thereby enhancing the overall fitness of the population.
2. Mutation: While crossover relies on the combination of existing traits within the population to generate new solutions, mutation introduces completely new genetic material into the population. During the mutation process, one or more genes in a solution are randomly altered to a different value. This operator is critical to maintain diversity within the population and to prevent premature convergence to sub-optimal solutions. Despite its crucial role, mutation is typically applied with a low probability to avoid excessive and disruptive changes to promising solutions.

Each of these operators plays a unique and critical role within the GA framework. Crossover and mutation stimulate diversity and innovation, respectively, while selection applies pressure to the population, promoting survival of the fittest solutions. It's the well-coordinated interplay among these operators that enables GA to effectively explore and exploit the solution space, driving the population towards optimality over successive generations.

a. Crossover

The cornerstone of genetic progression in GA, crossover, involves the ingenious fusion of two parent solutions to birth offspring solutions. This operator aims to incorporate superior traits from each parent to increase the overall fitness within the population.

Beginning with the selection of two parent solutions from the pool, a crossover point (or points in case of multiple-point crossover) is randomly selected along the length of their encoded chromosomes. Following this, segments of the parent chromosomes are exchanged at these points, producing a pair of new offspring chromosomes. In binary representation of chromosomes, this operation typically involves splitting the parent strings at the selected points, and then swapping the segments.

The strategy of selecting split points and the length of the exchanged segments (or genes) can greatly influence the effectiveness of the crossover operator and, subsequently, the final results. This strategy should be tailored based on the specific problem characteristics. An ideal crossover strategy fosters diversity, explores new solutions, and at the same time, preserves the dominant features of the parent solutions, thereby enhancing the population's fitness value.

Figure 3.10 schematically illustrates this process, showcasing how parent chromosomes are divided and then recombined to generate new offspring solutions. The intricate interplay of these processes encapsulates the essence of the crossover operator, contributing significantly to GA's ability to progressively evolve towards optimal solutions.

b. Mutation

As an integral part of GA, mutation introduces random modifications to existing solutions, aiming to foster diversity and broaden the search for an optimal solution. It operates on individual genes within a parent chromosome, generating offspring with slight alterations.

For binary-coded chromosomes, the mutation operator typically performs a bit-flip operation. In simpler terms, if a gene is coded as '0', it's mutated to '1', and vice versa, as outlined in the following function:

$$BinaryMutation(gen) = \begin{cases} 1 & \textit{if gen value} = 0 \\ 0 & \textit{if gen value} = 1 \end{cases}$$

Fig. 3.10 Typical crossover operator on a binary coded chromosome

Mutation plays a critical role in preventing premature convergence to local extremes by ensuring the GA doesn't become overly reliant on the currently dominant solutions. It effectively propels the algorithm to venture into potentially unexplored regions of the solution space, thus maintaining a healthy balance between exploitation of the current best solutions and exploration for new ones. This dynamic balance enhances GA's capability to locate the global optimum in complex problems.

Practical experience underscores the importance of tailoring the mutation function in accordance with the problem's specific characteristics. Overlooking this step can lead to the generation of non-viable offspring. By incorporating safeguards to prevent the algorithm from producing unfeasible solutions, one can significantly enhance the algorithm's overall performance.

3.12.5 Fitness Function

The fitness function, also known as an objective function, serves as the core decision-making mechanism within a Genetic Algorithm. It is designed to assess and rate the quality of individual solutions within the population. The aim of a GA is to find the solution that maximizes (or minimizes) the fitness function, depending on the specific problem at hand.

The fitness function provides a measure of suitability by assigning a fitness score to each individual in the population. These scores are then used to guide the selection of parents for the next generation. Individuals with higher fitness have a higher probability of being chosen for reproduction, thus ensuring the survival of the fittest solutions.

In formulating the fitness function, it's crucial to ensure that it accurately captures the objectives and constraints of the problem. This function must be able to quantitatively evaluate a solution's performance relative to the desired outcome.

In the case of TMD optimization, for example, the fitness function might be set to minimize the maximum displacement, velocity, and acceleration of a structure. The fitness of each solution (i.e., each set of TMD parameters) would then be calculated based on how effectively it reduces these measures.

However, designing an appropriate fitness function can be a complex task. It must strike a balance between encouraging exploration of the solution space (to find global optima) and exploitation of promising areas (to refine local optima). Inappropriate fitness functions may lead to premature convergence or failure to converge at all. By choosing the right fitness function, you set a course for the Genetic Algorithm, guiding it towards the optimal solution while avoiding pitfalls and dead ends. It's this pivotal role that makes the fitness function such a crucial part of the Genetic Algorithm.

3.12.6 Fast Elitist Multi-Objective Genetic Algorithm

In scenarios requiring simultaneous optimization of multiple conflicting objectives, such as in the optimization of TMDs, standard optimization methods fall short. Instead of a singular optimal solution, these multi-objective optimization problems entail a host of alternative solutions of equal merit, often termed as Pareto-optimal solutions.

The Fast Elitist Multi-Objective Genetic Algorithm II (NSGA-II) [94]. is a specialized algorithmic approach engineered to address these multi-objective optimization challenges. Its 'elitist' designation underscores its ability to preserve the superior individuals across generations, ensuring the survival of the fittest. Its 'fast' characteristic stems from its effective management and timely population updates.

NSGA-II implements a ranking method, assigning fitness values to the solutions based on their standing in Pareto dominance relations. In simple terms, a solution that dominates a larger subset of the population is deemed superior. To prevent early convergence to sub-optimal solutions and to support the exploration of the entire Pareto-optimal front, NSGA-II utilizes a niche formation method, promoting population diversity.

The NSGA-II algorithm begins with a randomly initialized population. Each individual in this population undergoes fitness evaluation and the entire population is ranked following the Pareto dominance principle. Next, the algorithm generates a new population via selection, crossover, and mutation operators, subsequently merging this with the original population. This unified population is then sorted based on both dominance rank and crowding distance, and the top individuals are selected to form the subsequent generation.

NSGA-II's ability to identify a diverse array of Pareto-optimal solutions in a single run affords it a distinct advantage over conventional optimization methods, which typically discover only one solution per run and often necessitate multiple iterations to uncover several Pareto-optimal solutions. Thus, NSGA-II emerges as a powerful tool in navigating complex multi-objective optimization challenges. In this section, we've delved into the intricacies of NSGA-II, enabling readers to gain an understanding of the algorithm's methodologies for ranking and sorting individuals within a population.

1. **Non-domination rank**

1. Non-Domination Ranking

In general terms, we can say that solution X dominates solution Y if X is not inferior to Y across all objectives, and if X outperforms Y in at least one objective. The next step involves identifying the non-dominated set within the population, which is then categorized as the *first front*. The 'second front' comprises the set of solutions that are only dominated by those in the first front. This stepwise process continues until all solutions within the population have been allocated to their respective 'fronts'. Subsequently, these fronts are sorted, starting with the first and ending with the last. Figure 3.11 provides a visual representation of solutions distributed across different ranks for a two-objective scenario [6].

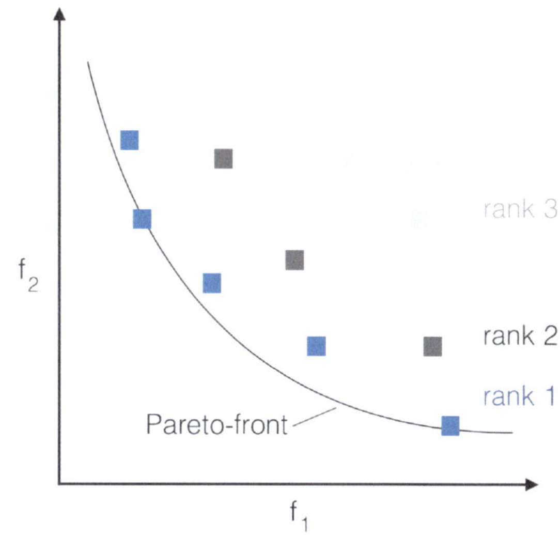

Fig. 3.11 Non-dominated sorting of solutions [6]

2. **Crowding distance**

For non-dominated solutions or a collective of the top-ranked non-dominated solutions, NSGA-II aims to achieve extensive coverage. This is facilitated through the use of crowding distance, defined as the Manhattan distance between adjacent solutions in a bi-objective scenario, as demonstrated in Fig. 3.12. In simpler terms, the algorithm attributes higher ranks to solutions that maintain a larger distance from their nearest counterparts. The reasoning behind this approach is the algorithm's intent to extensively explore the areas around these spaced-out solutions in the pursuit of superior, and ideally optimal, solutions.

Consequently, each solution that cannot be dominated by other solutions and has a larger crowding distance than the others will obtain the first rank and so on.

3.12.7 Repair Function

In the practical application of many problems, the crossover and mutation operators often generate numerous non-feasible solutions. This happens because these operators function mathematically on a binary string without considering the unique features and functions associated with each number in the context of the problem. Indeed, each segment of the chromosome pertains to and regulates a specific parameter.

There are two common strategies to address these non-feasible solutions. The first is by eliminating them from the population, and the second is by employing a repair function.

The repair function maps a solution from the infeasible region to a proximate solution in the feasible region. Figure 3.13 provides a schematic representation of the repair approach,

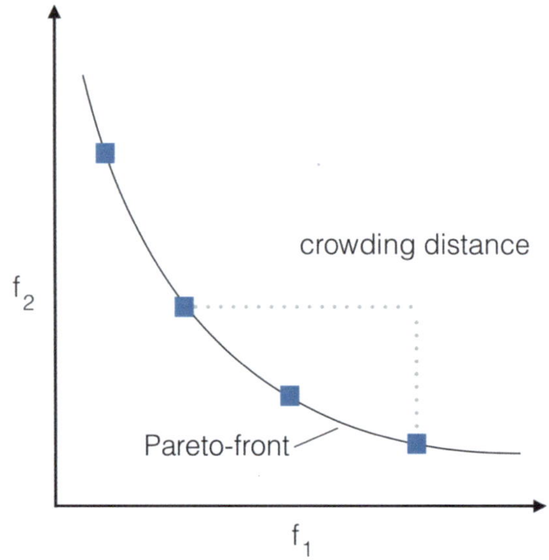

Fig. 3.12 Illustration of crowding distance [6]

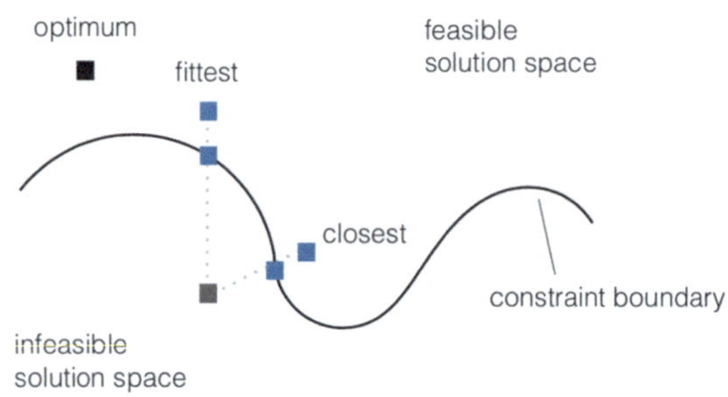

Fig. 3.13 Repair of an infeasible solution [6]

illustrating a solution space with an infeasible solution and two feasible solutions. Implementing a repair function within the GA framework offers distinct advantages, such as enhancing the performance of the algorithm by not overlooking potential optimums. The specific application of the repair function in the optimization of TMD parameters is further explored through various examples in this book.

Modeling and Design of TMDs 4

Having explored the theoretical aspects of Tuned Mass Dampers (TMDs), the focus now shifts from theory to application, implementing these potent vibration control devices in both research and practical engineering projects.

In the practical design and integration of TMDs, engineers frequently face multifaceted challenges related to precise modeling and efficient design realization. This chapter intends to not only highlight these challenges but also provide methodical, step-by-step guidance for modeling and designing TMD systems. The challenges are principally divided into three categories:

1. **Modeling**: Engineers generally begin with an existing Finite Element (FE) model of the structure using FE software. Integrating TMD(s) into this pre-existing model requires specific FE-modeling techniques, tailored for TMD systems. Such a strategy ensures that the TMDs are accurately embedded within the overall structural framework, thereby enabling a thorough analysis and design process.
2. **Initial Design**: This phase signifies the onset of enhancing structural behavior through the implementation of TMD systems. This section delves into the core principles guiding the initial placement and tuning of TMDs for achieving the targeted performance outcomes.
3. **Final Tuning**: The tuning phase is typically iterative and nuanced. The structural performance is continually assessed under assumed TMD parameters, and adjustments are made as needed. Engineers employ various performance indices to gauge and refine the control system's efficacy.

The intent of this chapter is to provide applicable methods to adeptly optimize TMD performance across diverse structural scenarios. Initially, techniques specific to TMD modeling

H. Radmard Rahmani, *Seismic Control of Structures Using Tuned Mass Dampers*,
Synthesis Lectures on Engineering, Science, and Technology,
https://doi.org/10.1007/978-3-031-71217-3_4

are articulated, followed by an introduction to two published benchmark problems for structural control systems. These benchmarks serve a crucial role, enabling researchers to test control strategies on well-structured models and compare their findings with those of their peers.

Upon outlining the modeling techniques, attention is directed towards the design parameters of the TMDs. This discussion aims to familiarize readers with the pragmatic design procedure of TMDs, including essential considerations that bridge theory with real-world applications.

Lastly, the chapter details execution techniques and other intricacies to be considered during the final planning stages, rounding out the design process. To further elucidate these concepts and provide concrete examples, the succeeding chapter offers practical illustrations of the principles presented here.

Through a cohesive blend of theory, methodology, and practical insights, this chapter serves as a comprehensive guide for those engaged in the modeling, design, and implementation of TMDs, bridging gaps and offering a navigable path through the complex landscape of structural vibration control.

4.1 Modeling TMDs

4.1.1 Introduction

The finite element (FE) model of a structure should encapsulate important characteristics that authentically echo the physical attributes of the structure. However, replicating every detail of a real-world structure is impractical, and an experienced engineer understands the trade-offs involved in model simplification, ensuring safety and reliability even amid uncertainties.

A typical structural model includes essential components such as beams, columns, and shear walls. It may also comprise other elements like foundations, masonry or infilled walls, or connection details depending on the requirements. Considering these details usually corresponds to the type of analysis performed. For instance, certain elements may significantly influence non-linear analysis but may be negligible in linear ones. Furthermore, the decision of whether to opt for dynamic or static analysis would also impact the modeling strategy.

When modeling TMDs and performing dynamic analysis, the level of detail in the FE model becomes a subject of interest. The guiding principle here is to keep the model as straightforward as possible while incorporating crucial components that have a noticeable effect on the outcomes. Determining what is "noticeable" often requires engineering judgement and sometimes necessitates separate modeling and computations.

The choice to omit specific components from the FE model involves disregarding their mass, stiffness, and damping properties. However, we can also model these attributes using suitable structural elements such as springs, joint masses, or dampers, simplifying our model to an abstract level.

Factors influencing the detail level in FE-modeling include:

- Material of structural elements: Steel, concrete, etc.
- Type of analysis: Linear or non-linear
- Analysis method: Static or dynamic
- Purpose of the analysis: Behavioral investigation or detailed engineering

Among these, the purpose of the analysis typically has the most impact on the level of detail in the FE-modeling, often directing other factors like the analysis method.

Let's discuss the analysis purpose for structures equipped with TMDs. Commonly, the projects involve the following phases:

Behavior Investigation:

This phase typically involves studying the structure's behavior under external excitations, both with and without control systems. The goal could be the final investigation in research projects or part of the decision-making process for control systems in engineering projects. During this stage, the main structural elements—columns, main girders, and the lateral load bearing system—as well as a simple model of the TMDs should be incorporated. Effects of secondary structural elements should be safely accounted for, including their mass, even if damping and stiffness effects are simplified. This phase will also focus on satisfying critical design criteria related to stability, serviceability, and durability.

For structures with multiple TMDs, optimizing their arrangement is an important goal. By the end of this phase, we should have robust estimates of the main structural elements' dimensions and specifications, the behavior under dynamic loads, and cost and time projections for project management.

Detailed Engineering:

This phase occurs during the project's advanced stages when the control system's primary design and main structural elements have been finalized. Here, secondary structural elements and finer details are designed, and final tuning of TMD parameters is carried out.

There are two main strategies during this phase. The first approach involves adding most secondary components to the FE-model and designing them concurrently with the main elements. The second strategy advises including only those secondary elements that necessitate complex modeling and can be modeled using simple elements. Some of these elements can add degrees of freedom to the model and should be included only when their impact is significant.

Also, during this phase, any modifications in architectural and structural plans may occur, necessitating close cooperation between architects and structural engineers. Although commercial software solutions make it easier to add details to the model and validate them against the code, it could lead to complex FE models that require advanced modeling skills. So, based on practical experience, designing structural details like connections is often done separately, using the results from the main structural model. Even when details aren't explicitly modeled, their influences must be carefully considered.

For example, the beam-column connection details can be accommodated in the model by modifying the connection rigidity in linear analysis. In contrast, non-linear analysis may consider the effect of connection details using non-linear elements and corresponding force-displacement diagrams based on structural codes like FEMA-356. It's worth mentioning that when using TMDs, it's usually undesirable to allow structural elements to enter a non-linear state.

In this section, we present modeling techniques for structures equipped with TMDs, beneficial for both researchers and engineers in the above-discussed scenarios.

4.2 Benchmark Problem Explained

4.2.1 Introduction

This section can be in particular beneficial for the researchers and engineers that want to test their control methods on a structure and looking for a standard structural model that its structural parameters is available and is well designed based on the building codes. The goal of this section is to familiar the reader with a benchmark problem which are specifically developed for structural control problems. All the structural parameters of the benchmark problem are available online. Moreover as it is already used in many research projects on the structural control system, the performance of the developed controller can be compared with the others which gives a better view.

In research applications, the emphasis often lies more on behavioral investigation than on detailed design. Typically, a 2D model of the structure is prepared, reflecting the behavior of the main lateral load-bearing system of the building. If a 3D model of the structure exists, an abstracted 2D-model may be built by neglecting elements that do not contribute to lateral load-bearing. Degrees of freedom (DOFs) reduction techniques, such as static condensation or the Ritz method, are then employed.

The objective of such simplifications is to minimize the complexity of the system, thus facilitating easier interpretation of results and reducing computational costs. In this context, a code for the dynamic analysis of structures using the Newmark-β method is presented within this book, which is particularly suitable for research applications.

To model the TMD in a 2D FE-model, an additional horizontal DOF is added to the structure for each TMD, and the stiffness, mass, and damping values of the TMDs are incorporated into the corresponding matrices in those DOFs. A typical process for preparing structural matrices in a research project is outlined as follows:

1. **Determine Structural Matrices**: The structural matrices $[M]$ and $[K]$ for the structure without TMD are first determined, representing the mass and stiffness respectively.
2. **Divide DOFs**: The DOFs are categorized into static and dynamic groups.
3. **Reduce Matrices**: The structural matrices are reduced by eliminating most of the static DOFs using the static condensation method. This step is critical to focus the analysis on the dynamics of the system.
4. **Calculate Damping**: Matrix: The damping matrix, $[C]$, is calculated for the reduced FE model. Rayleigh damping, as described in the previous chapter, is typically used for this purpose.
5. **Incorporate TMD Parameters**: The TMD's parameters are integrated into the structural matrices by defining new DOFs and extending the structural matrices. This step is vital to capture the influence of the TMDs on the system's behavior.

To elucidate the process, we can review a practical example: the reduction of DOFs in a 75-story benchmark building, cited as [95] . This example has been widely used in structural control research projects.

The application of these steps effectively encapsulates the critical elements of the structure and the behavior of the TMDs. It reflects a balance between representing essential characteristics and maintaining a manageable level of complexity, allowing for accurate yet efficient analysis.

4.2.2 Model Review: Benchmark Control Problem for Seismically Excited Buildings

A widely utilized benchmark building in seismic control problems is presents the problem definition and guidelines of the next generation structural control benchmark problem for seismically excited buildings. Focusing on a 20-story steel structure representing a typical mid- to high-rise building designed for the Los Angeles, California region, the goal of this study is to provide a clear basis to evaluate the effect of various structural control strategies. An evaluation model has been developed that portrays the salient features of the structural system. Control constraints and evaluation criteria are presented for the design problem.The task of each participantin this benchmark study is to define (including devices, sensors and control algorithms), evaluate and report on their proposed control strategies. These strategies may be either passive, active, semi-active or a combination of them. The structural parameters of this building is prepared and published by Spencer et al. [7].

The Los Angeles twenty-story (LA 20-story) structure is 30.48 m (100 ft) by 36.58 m (120 ft) in plan, and 80.77 m (265 ft) in elevation. The bays are 6.10 m (20 ft) on center, in both directions, with five bays in the north-south (N-S) direction and six bays in the east-west (E-W) direction. The building's lateral load-resisting system is comprised of steel perimeter moment-resisting frames (MRFs). The interior bays of the structure contain simple framing with composite floors.Th building has five bays in north-south (N-S) and six bays in east-west directions. The lateral load bearing system of the building comprises of steel perimeter moment frames (MRFs). The authors have developed an in-plane finite element model of the N-S frame for research purposes. The initial model of the building was a 2D-frame with 180 modes which results in 540 DOFs (two transitional and one rotational DOF) as it is shown in Fig. 4.1. The DOFs corresponding to fixed boundary conditions are then constrained by eliminating the rows and columns associated with these DOFs from the global mass and stiffness matrices. The constrained DOFs are the horizontal DOFs at nodes 1, 2, 3, 4, 5, 6, 13 and 18 and the vertical DOFs at nodes 1, 2, 3, 4, 5 and 6. Realizing these 14 boundary conditions results in the mass matrix, M [526 × 526], and the stiffness matrix, K [526 × 526].

DOFs Reduction Using Ritz Transformation

Because the floor slab is assumed to be rigid in its horizontal plane, the nodes associated with each floor have the same horizontal displacements. This assumption is enforced by writing constraint equations relating the dependent (slave) horizontal DOFs on each floor slab to a single active horizontal DOF and using a Ritz transformation. First, the structural responses are partitioned in terms of active and dependent DOFs as $U = [U_a U_d]$, and the constraint equations are written in the form:

$$K_{da}U_a + K_{dd}U_d = \mathbf{0} \tag{4.1}$$

Here, K_{da} and K_{dd} are parts of the stiffness matrix corresponding to active (independent) and dependent (slave) DOFs, respectively. U_a and U_d are the corresponding displacement vectors for the active and dependent degrees of freedom.

The mass and stiffness matrices are similarly partitioned in terms of active and dependent DOFs:

$$[M] = \begin{bmatrix} M_{aa} & M_{ad} \\ M_{da} & M_{dd} \end{bmatrix} \tag{4.2}$$

$$[K] = \begin{bmatrix} K_{aa} & K_{ad} \\ K_{da} & K_{dd} \end{bmatrix} \tag{4.3}$$

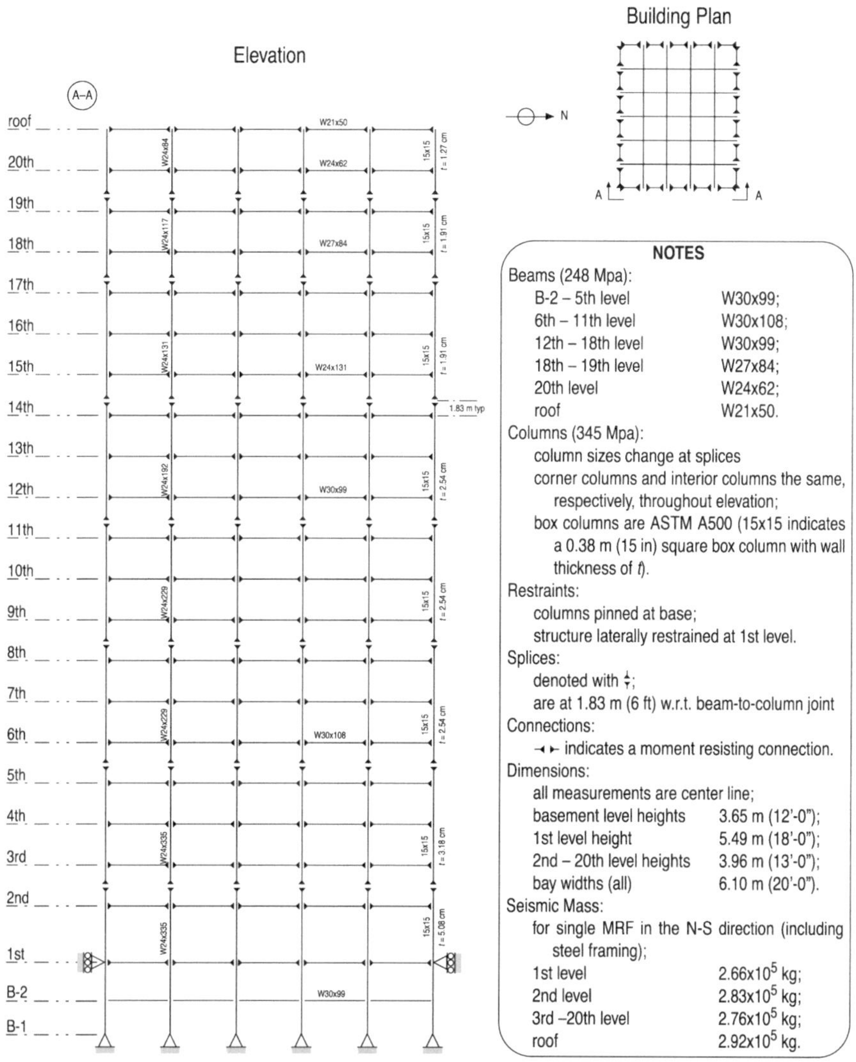

Fig. 4.1 Benchmark control problem for seismically excited buildings: Los Angeles 20-story building N-S MRF [7]

The equation of motion for the undamped system, once the dependent degrees of freedom have been reduced, is given by:

$$\left[\hat{M}\right]\{\ddot{U}\} + \left[\hat{K}\right]\{U\} = -\left[\hat{G}\right]\ddot{u}_g \tag{4.4}$$

Where $\left[\hat{M}\right]$ and $\left[\hat{K}\right]$ are reduced matrices and are calculated as follow:

$$\left[\hat{M}\right] = \left[T_R^T\right][M][T_R] \tag{4.5}$$

$$\left[\hat{K}\right] = \left[T_R^T\right][K][T_R] \tag{4.6}$$

$$\left[\hat{G}\right] = \left[T_R^T\right][M][\Gamma] \tag{4.7}$$

Where:

$$T_R = \begin{Bmatrix} I \\ T_{da} \end{Bmatrix} \tag{4.8}$$

$$T_{da} = -R_{dd}^{-1} R_{da} \tag{4.9}$$

Explanation of the terms:

- T_R is the transformation matrix used to link the original DOFs (both active and dependent) with the reduced set of DOFs.
- The partitioning of the mass and stiffness matrices into active a and dependent$d$ components corresponds to the partitioning of the displacement vector.
- The matrix I is the identity matrix corresponding to the active DOFs, meaning that active DOFs remain unchanged in the transformation.
- T_{da} is a sub-matrix of T_R, and it is defined as the negative inverse of R_{dd} multiplied by R_{da}. It describes the transformation relationship between dependent and active DOFs, enforcing the constraints that were described earlier.
- Γ is a vector of zeros and ones defining the loading of the ground acceleration to the structure.
- Applying the transformationT_R to the mass and stiffness matrices results in the reduced mass $\left[\hat{M}\right]$ and stiffness $\left[\hat{K}\right]$matrices, which describe the system in the reduced set of DOFs.

By utilizing the Ritz transformation and reducing the horizontal dependent DOFs the number of the DOFs is reduced to 418.

DOFs Reduction Using Guyan

Due to the complexity and sheer size of the model, it is imperative to reduce the number of DOFs to a more manageable scale without sacrificing the essential dynamics of the full model. This can be achieved through Guyan reduction, a technique employed to maintain the main dynamic characteristics of the system while simplifying the model. Utilizing the reduced mass and stiffness matrices obtained through Guyan reduction, the damping matrix is then established on the basis of an assumption of modal damping, allowing the formulation of the equations of motion for the structure. These equations are structured in a state-space form, making them suitable for detailed design and analysis.

In this specific model, the natural frequencies associated with higher modes appear to be large, an occurrence attributed mainly to rotational and vertical DOFs. Considering that these modes are unlikely to influence the physical system's response, they are deemed unnecessary for the benchmark model and can thus be reduced or eliminated. Implementing Guyan reduction once again, all rotational and most vertical DOFs are condensed, scaling down the original 418 DOF model to nearly one-fifth of its initial size. This reduction focuses on separating the DOFs into active (independent) and dependent (slave) categories. The active horizontal DOFs, along with vertical DOFs for levels 1–21 strategically located on the second and fifth column lines, are designated as active. In contrast, all remaining vertical DOFs, including those at splice locations, and all rotational DOFs are deemed dependent and effectively condensed out of the model (Fig. 4.2). The resulting mass and stiffness matrices, partitioned to reflect the distinction between active and dependent degrees of freedom, are thus established, setting the stage for further analysis and application.

$$\left[\hat{M}\right] = \begin{bmatrix} \hat{M}_{aa} & \hat{M}_{ad} \\ \hat{M}_{da} & \hat{M}_{dd} \end{bmatrix} \tag{4.10}$$

$$\left[\hat{K}\right] = \begin{bmatrix} \hat{K}_{aa} & \hat{K}_{ad} \\ \hat{K}_{da} & \hat{K}_{dd} \end{bmatrix} \tag{4.11}$$

The equation of motion for the undamped system, once the dependent degrees of freedom have been reduced, is given by:

$$\left[\tilde{M}\right]\{\ddot{U}\} + \left[\tilde{K}\right]\{U\} = -\left[\tilde{G}\right]\ddot{u}_g \tag{4.12}$$

where $\left[\hat{M}\right]$ and $\left[\hat{K}\right]$ are reduced matrices and are calculated as follow:

$$\left[\tilde{M}\right] = \left[T_G^T\right]\left[\hat{M}\right][T_G] \tag{4.13}$$

$$\left[\tilde{K}\right] = \left[T_G^T\right]\left[\hat{K}\right][T_G] \tag{4.14}$$

where:

$$T_G = \begin{Bmatrix} I \\ \hat{T}_{da} \end{Bmatrix} \tag{4.15}$$

$$\hat{T}_{da} = -\hat{K}_{dd}^{-1}\hat{K}_{da} \tag{4.16}$$

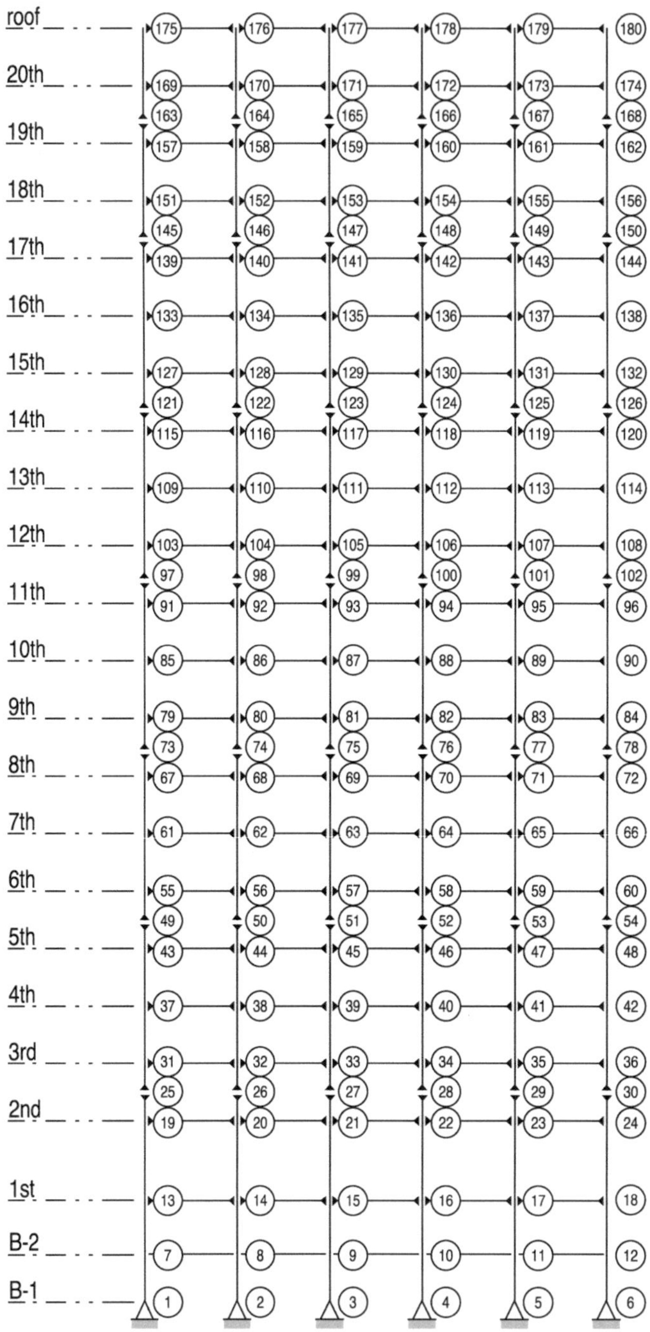

Fig. 4.2 DOF number in LA-20 story building

The damping matrix C is define based on the reduced system and the assumption of modal damping. Damping in each mode is assumed to be proportional to the mode's associated frequency, with a maximum of 10% critical damping in any one mode. The damping in the firs mode is assumed to be 2%; therefore the damping ζ_i in the ith mode is given by:

$$\zeta_i = min\left\{\frac{\omega_i}{50\omega_i}, 0.1\right\} \tag{4.17}$$

where ω_i is the natural frequency of the ith mode. The damping matrix, C is then determined via:

$$\tilde{C} = \tilde{M}\mathbf{\Phi}\begin{bmatrix} 2\xi_i\omega_i & 0 & 0 \\ 0 & \dots & 0 \\ 0 & 0 & 2\xi_n\omega_n \end{bmatrix}\mathbf{\Phi}^{-1} \tag{4.18}$$

where $\mathbf{\Phi}$ is the matrix of mode shapes.

The Guyan reduction results in a final model with 106 DOFs, that maintains the important dynamics of the original model. The first three natural modes of the structure is shown in the Fig. 4.3.

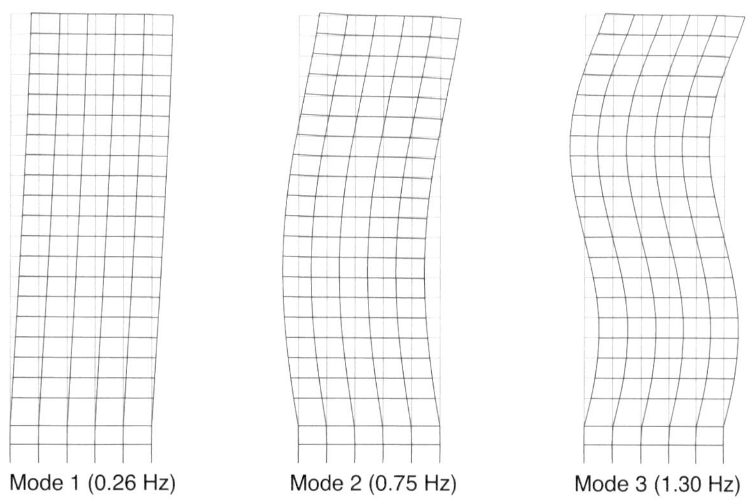

Fig. 4.3 DOF number in LA-20 story building

Summary of Insights Gained from the Benchmark Problem

The benchmark problem has provided valuable insights and resources that serve to advance the understanding and application of structural control systems. The specific contributions and learning outcomes derived from this problem include:

1. **Structural Parameters Availability**: The detailed structural parameters for a high-rise building are readily available, offering a comprehensive foundation for testing various control algorithms, enhancing the accessibility and replicability of the experiments.
2. **Simulink Model Accessibility**: A Simulink model of the structure has been prepared, greatly facilitating the implementation of control strategies. This readily accessible model streamlines the development and testing process, allowing for more efficient exploration of different control mechanisms.
3. **Performance Evaluation Platform**: The established benchmark enables the performance of controller systems to be evaluated in a systematic and standardized manner. By comparing results with other published findings, researchers can gauge the effectiveness and robustness of their control strategies, contributing to a shared knowledge base.
4. **Model Reduction Techniques**: The techniques utilized for model reduction within the benchmark problem are thoroughly explained, providing valuable methodologies that can be applied to other models. These techniques not only offer a pathway to simplification but also maintain the critical dynamics of the system, enhancing their broader applicability.
5. **Framework Testing and Comparison**: The model serves as a versatile tool that can be used as input to check newly developed frameworks. By enabling direct comparison with existing results, it fosters consistency and continuity in the research community, driving innovation while upholding rigorous standards.

Through these key takeaways, the benchmark problem contributes significantly to the field, enriching both the theoretical understanding and practical application of structural control systems in high-rise buildings. Its multifaceted contributions underscore the importance of well-designed benchmarks in advancing research and shaping future innovations.

4.3 Designing the TMD Parameters

This section serves as a practical guide for designing a TMD system for structures exposed to external dynamic loads. While abundant methods and solutions are available in academic literature, navigating them all can be daunting. Therefore, this section aims to offer a succinct yet comprehensive overview, distilling essential theory and practice to facilitate straightforward application.

The guide is structured into three main components for ease of understanding:

Key Aspects: This introductory portion provides a holistic view of the entire design process to give readers a complete picture of what to expect.

Number and Position of TMDs: This segment initiates the design guide, focusing on determining the ideal number and locations for the TMDs within the structure.

Optimization of TMD Parameters: Finally, we delve into fine-tuning the specific parameters of the TMD system, starting with recommended mass considerations and proceeding to optimal tuning frequency and damping ratio. To illustrate the workflow, a flowchart will be presented at the end of this section. Additionally, practical examples will be provided in the subsequent chapter to further clarify the approach outlined here.

4.3.1 Key Aspects

The effectiveness of a TMD in controlling unwanted vibrations is highly dependent on its proper tuning to the specific dynamics of the system. Below, the key aspects and steps involved in designing the TMDs for structural control are listed. Which are then explained in the further sections.

1. **Determination of the External Excitation**
   - Recognizing the possible dynamic loads on the structure which can be classified in two main categories: Unharmonic loads such as earthquake, wind, and harmonic loads.
   - Preparing the external excitation record by selecting and scaling the records based on the location of the structure and related building codes.
2. **Determining of the dynamic characteristics of the structure**
   - Natural Frequency Analysis: Identifying the structure's natural frequencies, especially the primary modes that are to be controlled.
   - Mode Shapes: Examining the mode shapes to understand the modal deformations for different structural modes.
   - Response Analysis: Analyzing the expected structural response to the external excitations such as wind or earthquake loading.
3. **Selection of TMD Type**
   - Single or Multiple Mode control: Deciding whether a single or multiple modes are required to be controlled based on the structural responses and type and characteristics of external excitation.
   - Type of Damper: Selecting the appropriate damper type (e.g., linear, nonlinear) based on the system's requirements.

4. **TMD Parameter Determination**

   - Mass Ratio (μ): Determine the mass ratio of the TMD to the structure.
   - Damping Ratio (ξ): Determine the optimal damping ratio to achieve desired control without overdamping the system.
   - Tuning Frequency (ν): Calculate the frequency at which the TMD should be tuned. Often, it is close to or equal to the natural frequency of the mode to be controlled.

5. **Validation and Adjustment**

   - Simulation and Testing: Validate the selected parameters through simulation and experimental testing, if possible.
   - Iterative Tuning: Adjust the parameters iteratively, as necessary, to optimize the system's performance under various scenarios.
   - Consideration of Uncertainties: Evaluate the robustness of the TMD system against uncertainties in the structure's properties or loading conditions.

6. **Integration with Overall Design**

   - Physical Implementation Considerations: Address practical constraints such as available space, weight limitations, and maintenance requirements.
   - Interaction with Other Systems: Assess the interaction of the TMD with other structural components and systems to ensure compatibility.

Conclusion

The determination of TMD parameters is a challenging task that requires theoretical comprehension and practical insights. Before embarking on the calculation of these parameters, data concerning the structural modes must be attained. This is generally accomplished through a modal analysis performed on the Finite Element (FE) model of the uncontrolled structure.

The base model considered for designing the TMD system must be the final design model, meaning that all structural elements must meet code criteria for stability and cross-section resistance. In this model, the weight of the TMDs and the forces they generate, including potential stroke forces, are taken into account. As an estimation, two times the TMD weight may be employed as a horizontal force to consider local effects on structural members.

The dynamic effects are subsequently considered using dynamic analysis, and there may be a need to strengthen structural elements based on these results. This process demands an understanding of modal analysis theories, followed by the utilization of specific design processes to determine the TMD parameters.

While various approaches exist for calculating TMD parameters, certain universal principles and specifications must be adhered to in the design process. These rules provide foundational guidance to ensure that the design aligns with the underlying structural requirements and performance expectations. The objective of this section has been to provide a guide for the proper determination of the TMD parameters.

4.3.2 Determination of the Position of the TMD

Determining the optimal position of a TMD is a crucial step in the design and implementation of structural control systems. The TMD's position within a structure can significantly impact its effectiveness in mitigating vibrations and controlling structural response. Here's how this complex process is typically approached:

1. **Understanding the Structural Behavior**: The first step involves a detailed understanding of the structure's behavior, especially the modes that are most likely to be excited by external forces such as wind or earthquakes. The identification of critical areas or floors where vibrations are most pronounced can guide the positioning of the TMD.
2. **Assessing Structural Constraints**: Physical constraints within the structure, such as available space, architectural considerations, and connection possibilities, must be assessed. These constraints may limit or guide the placement options for the TMD.
3. **Analyzing Modal Characteristics**: The natural frequencies and mode shapes of the structure must be analyzed to find the optimal location for the TMD. This usually involves a comprehensive modal analysis to understand how the TMD's position might influence different modes of vibration.
4. **Optimizing for Effectiveness and Efficiency**: The optimal positioning of the TMD will consider not only its effectiveness in controlling vibrations but also its efficiency in terms of material usage, ease of installation, and maintenance considerations. Optimization techniques can be applied to balance these factors.
5. **Integration with Structural Elements**: Consideration must be given to how the TMD will be integrated with existing structural elements. The position must allow for the proper functioning of the TMD while also respecting the structural integrity and load paths of the existing structure.
6. **Compliance with Regulations**: Compliance with relevant building codes and regulations is mandatory. This may include considerations related to safety, accessibility, and other legal requirements that could influence the positioning of the TMD.
7. **Consideration of Future Adjustments**: A forward-thinking approach is necessary, keeping in mind potential future adjustments, modifications, or expansions. This

ensures that the chosen position for the TMD remains viable and effective throughout the lifespan of the structure.

From a control engineering standpoint, the optimal location for a single TMD is typically the story experiencing the maximum modal displacement at the control mode with lowest frequency. In most buildings, this is often the topmost story. However, there are exceptions like in tall buildings where positioning the TMD on other floors can enhance the overall performance of the structure when subjected to external forces.

4.3.3 Number of TMDs

This section delves into the crucial decision-making process surrounding the quantity of TMDs to be deployed in a structure. The number of TMDs is not just a question of maximizing damping capabilities; it also has significant ramifications for structural integrity, economic feasibility, and architectural integration. While a single TMD may suffice for simpler structures with straightforward dynamic behaviors, more complex buildings—especially tall or irregularly shaped ones—may necessitate multiple TMDs to adequately control multiple vibration modes. We will explore the factors that guide this pivotal decision, laying out scenarios where a single TMD may be adequate and instances where multiple TMDs become indispensable.

Single TMD

In the context of mid- to high-rise buildings governed by seismic codes, the control of initial modes by a TMD system is a conventional approach. This stems from the seismic specifications embedded within these codes, steering structural designs towards a shear-building form. For instance, code regulations deter the creation of weak stories by enforcing standards on the relative stiffness of each story in comparison to the one above it. Technically, in shear building forms, initial modes relate to the cantilever deformation of the structure in both X and Y directions, as well as a rotational mode shape around the center of stiffness.

A common strategy when utilizing a single TMD to control one of the natural modes in a direction is to place the TMD near the building's planar center of stiffness. A key metric for gauging the participation of each mode in the structure's overall response is the mass participation ratio. This ratio elucidates the percentage of mass activated in a particular mode, reflecting its contribution during earthquake excitation.

In some instances, rotational modes exhibit significant mass participation ratios. In these situations, controlling modes with rotational displacements can enhance the overall response of the structure. For such cases, two approaches are generally considered:

1. **Modify Structural Elements**: Without TMD, examine if rotational modes can be managed by either reducing rotational moments or improving stiffness in these OOFS. This can be achieved by relocating main lateral load-bearing systems, amending architectural plans to permit more balanced mass and stiffness distribution, or shifting lateral load-bearing systems towards the structure's perimeters. This latter alteration often positively impacts the control of rotational effects.
2. **Utilize Dual TMDs**: Consider the strategic placement of two TMDs, positioned symmetrically away from the center of mass. By leveraging the eccentricity of the control forces, the rotational effects can be reduced. While it is generally preferred to minimize the number of TMDs for architectural and financial reasons, certain scenarios favor multiple TMDs to optimize control, as elaborated in subsequent sections.

While minimizing the number of TMDs in a structure is often favored for architectural harmony and financial efficiency, certain circumstances necessitate the employment of multiple TMDs. These specific scenarios, where multiple TMDs offer distinct advantages, are expressed in the following section.

> It is essential to recognize that during an earthquake, the excitation will act on the masses of the structure, causing resultant forces to be applied at the center of mass. These forces are then transferred and carried by the structural lateral load-bearing system at the center of stiffness. Any discrepancy between these two centers can lead to rotational forces. Specifically, the greater the distance between the center of mass and the center of stiffness, the more substantial the rotational forces that will be produced. This alignment consideration is crucial in the design and analysis of structures to ensure that they can adequately withstand the dynamic forces induced by seismic events.

Multiple TMDs

The utilization of multiple TMDs often provides a wider control range for the control system, an essential feature when a structure possesses more than one natural frequency with significant participation in response to external excitations. Given the broad frequency content typical of most earthquakes, this situation may arise during seismic events, making multiple TMDs a viable solution.

Multiple TMDs, tuned to different frequencies, can be strategically placed at locations of maximum structural modal displacement for modes close to those frequencies. This approach extends beyond simple placement, with more complex distributed configurations potentially offering optimal solutions for tall buildings.

In case of tall buildings, the situation may become more complex, with the participation of higher modes significantly influencing the total response of the structure. Recent research underscores the potential benefits of controlling higher modes in tall buildings, as it can notably enhance the performance of control systems.

One method of achieving control over multiple modes is to employ multiple TMDs. These can be either positioned within a single story or distributed along the height of the structure. Rahmani et al. [96] have demonstrated that distributed TMDs can enhance the control system's performance in tall buildings when subjected to earthquake excitations. Further studies have corroborated these findings, highlighting the advantages of using distributed systems in comparison to a single TMD for controlling vibrations in high-rise buildings under the influence of both wind and earthquake loads [34, 50, 97, 98].

This real-world example highlights the challenging task of providing adequate space in the architectural plan for the TMD, considering the amplitude of its movement during external excitations. Care must be taken to position the supports of the TMD in the center of the stiffness of the story, thereby reducing the unwanted torsional moment resulting from the control forces multiplied by the distance between the supporting point and the center of stiffness of the story.

The selection of the optimal position contributes substantially to the effectiveness of the TMD, making this a pivotal aspect of the design and implementation process. The goal is to ensure that the chosen position for the TMD remains viable and effective throughout the lifespan of the structure, while also accommodating architectural considerations and practical constraints.

4.3.4 Determination of TMD Mass

The design of a TMD for vibration mitigation requires careful consideration of its mass, as the mass plays a important role in the TMD's capacity to apply forces opposite to those induced by an earthquake. Among the three design parameters of the TMD—mass, damping, and stiffness – mass is the first aspect to be considered, yet it is the last parameter we wish to modify or increase, due to its substantial impact on the overall system.

During external excitations, the TMD's oscillation off-phase enables it to counteract the structure's response to seismic forces. By employing a larger mass in the TMD, the capacity of the control system can be increased in terms of the magnitude of lateral control forces. However, the selection of mass is constrained by several key factors:

- **Structural Provisions**: A larger TMD mass generates higher internal forces in the structural elements that bear the TMD's load, and it also produces more substantial strokes that must be accommodated by the associated structural elements. These factors limit the feasible mass of the TMD.

- **Architectural Provisions**: Greater TMD mass necessitates more space within the architectural plans, which may conflict with design intentions and functional requirements.

Considerations in Mass Determination

There are several factors that should be considered in determination of TMD mass:

1. Active Mass Determination: This includes the dead load of the structure and a percentage of the live loads, usually between 20% to 30%. Special cases should also be evaluated for stability and serviceability, including scenarios with maximum live load and zero live load, as they represent extraordinary situations for TMD design.
2. Structural Loading: The mass must be balanced against the additional stresses it imposes on the structure's elements.
3. Economic and Architectural Considerations: Cost, placement, and aesthetics must be balanced against the mass's control benefits.
4. Type of External Forces: Wind, seismic forces, and other conditions must be taken into account to calibrate the mass for effectiveness under specific conditions.
5. Multiple Modes Control: If the TMD aims to control multiple vibration modes, the mass must be selected accordingly.
6. Materials and Construction: The mass can also be influenced by the materials used and construction techniques employed.

While numerous studies have explored the optimization of TMD mass, their suggested "optimal" values often exceed what is practically feasible for real-world applications. In practice, the mass of the TMD is usually determined as a fraction of the structure's total active mass, typically ranging between 1% and 3%. This approach balances theoretical effectiveness with practical constraints, offering a more applicable guideline for engineers and designers.

$$\mu = \frac{m_{st}}{m_t} \tag{4.19}$$

The active mass encompasses the parts of the structure that are susceptible to horizontal accelerations during external excitations, such as earthquakes. According to some seismic codes, this mass is calculated by combining the dead loads with approximately 30% of the total live loads. However, the determination of the active mass is not merely a matter of following a standard formula. It requires careful judgment by the engineer, who must consider the specific characteristics and functional requirements of the building.

The initial step in determining the mass for a TMD involves assessing the structure's maximum capacity to bear vertical loads and stroke forces, which the TMD might subsequently impose on the structure. These calculated values establish the boundary conditions or limit lines for the TMD design, ensuring that the chosen mass aligns with the structural integrity and constraints, and does not exceed the structure's ability to safely absorb the forces generated by the TMD.

In the practical determination of TMD mass, an iterative process is typically employed. Of the three main parameters of TMD, mass has the most significant impact on other structural parts, as an increase in TMD mass produces greater loads on both the vertical and horizontal load-bearing systems. Consequently, it is preferable to initially seek optimal TMD performance by adjusting the damping and stiffness rather than the mass. The process often begins by tuning the TMD with a conservative mass ratio, such as 1%, and then calculating and evaluating structural responses and stroke forces against predetermined acceptable standards. By manipulating the damping and tuning frequency of the TMD, the maximum possible performance at this mass ratio is assessed. If this performance does not satisfy the established sustainability and serviceability criteria, the mass of the TMD may be increased and other parameters are optimized based on the new mass value, to explore potential improvements in system performance. This iterative approach continues, with ongoing analysis of structural responses shaping successive adjustments, until the optimal performance is realized, all within the acceptable constraints for the TMD's mass.

4.3.5 Optimal Tuning Frequency and Damping Ratio

After selecting an appropriate mass for the TMD, the next critical step is tuning its frequency to match the structural natural frequency that needs to be controlled. The subject of identifying the optimal tuning frequency for TMDs has been extensively studied in the past, and numerous research articles address this issue [29, 33, 34, 41, 42, 44, 55, 66, 96, 99, 100]. One foundational reference that has been widely used over the years for tuning TMDs is the set of guidelines recommended by Warburton [101].

These recommended optimum values are presented in Table 4.1, where α represents the frequency ratio, ζ signifies the damping ratio, and μ denotes the mass ratio between the TMD and the structure. Subscripts t and st are used to differentiate between the TMD and the structure, respectively.

$$\alpha = \frac{\omega_t}{\omega_{st}} \tag{4.20}$$

$$\zeta = \frac{c_t}{c_{cr}} = \frac{c_t}{2m_t\omega_t} \tag{4.21}$$

$$\mu = \frac{m_t}{m_{st}} \tag{4.22}$$

Table 4.1 Optimal TMD parameters

	External excitation		Optimal TMD parameters		
	Type	Applied to	Optimized for	α_{opt}	ζ_{opt}
1	Force	Structure	X	$1+\frac{2}{\mu}$	$\sqrt{\frac{3\mu}{8(1+\mu)}}$
2	Force	Structure	$\ddot{X}$	$\sqrt{1+\frac{2}{\mu}}$	$\sqrt{\frac{3\mu}{8(1+\frac{\mu}{2})}}$
3	Acceleration	Base	X	$\frac{\sqrt{1-\frac{\mu}{2}}}{1+\mu}$	$\sqrt{\frac{3\mu}{8(1+\frac{\mu}{2})(1-\frac{\mu}{2})}}$
4	Acceleration	Base	$\ddot{X}$	$\frac{1}{1+\mu}$	$\sqrt{\frac{3\mu}{8(1+\mu)}}$
5	Random force	Structure	X	$\frac{\sqrt{1+\frac{\mu}{2}}}{1+\mu}$	$\sqrt{\frac{\mu(1+\frac{3\mu}{4})}{4(1+\mu)(1+\frac{\mu}{2})}}$
6	Random acceleration	Base	X	$\frac{\sqrt{1-\frac{\mu}{2}}}{1+\mu}$	$\sqrt{\frac{\mu(1-\frac{\mu}{4})}{4(1+\mu)(1-\frac{\mu}{2})}}$

The recommended values for TMD tuning are categorized based on four principal types of external excitations: force, acceleration, random forces, and random acceleration. The first two are pertinent to harmonic excitations, while the latter two concern stochastic loads such as winds and earthquakes. The optimization process accounts for both displacements and accelerations, which are key metrics for evaluating structural behavior. It's important to note that the initial values in Table 4.1 were investigated in the context of undamped systems. Subsequent research by Ioi and Ikeda [102] has demonstrated that the influence of system damping on optimal TMD parameters is less than 1% for $0.03 \leq \mu \leq 0.40$ and $0 < \zeta_t \leq 0.15$. Further studies by Warburton [103] indicate that these optimal values are applicable to both SDOF and Multiple MDOF systems.

Stiffness determination of TMD

Once the optimal parameters have been identified from the table, the optimal stiffness for the TMD can subsequently be calculated using the equations provided below:

$$k_t = m_t \alpha_{opr}^2 \omega_{st}^2 \tag{4.23}$$

where:

ω_{st} : Angular frequency of the structure
k_t : Stiffness of the TMD
m_t : Mass of the TMD

The stiffness parameter essentially governs how the mass within the TMD moves in relation to the structure it is attached to. This relationship plays important role in TMD's performance; an improperly tuned stiffness parameter could lead to ineffective control of structural vibrations or even exacerbate them in certain scenarios. The ideal stiffness ensures that the TMD operates in a way that offsets the disruptive forces acting on the structure, thereby enhancing both its safety and functional integrity.

Damping of the TMD

Using the optimal damping ration in Table 4.1, the mass and the stiffnesses values, the TMD's damping can be calculated as follow:

$$c_t = 2\zeta_{opt} m_t \sqrt{\frac{k_t}{m_t}} \tag{4.24}$$

Understanding the influence of the damping ratio on the performance of a TMD is crucial for effective design. Contrary to common belief, increasing the damping ratio does not always lead to improved performance. To shed light on this nuance, this section presents two key studies that explore the effects of damping ratios in distinct contexts.

The inaugural study is by Den Hartog, published in 1956 [104], which investigates the amplification factor for a SDOF system equipped with an undamped TMD, subjected to harmonic loading. In the context of sinusoidal excitations, Den Hartog derives the dynamic amplification factor (R) as follows:

if:

$$P = P_0 sin(\omega t) \tag{4.25}$$

then:

$$R = \frac{U_0}{U_{st}} = \sqrt{\frac{(\alpha^2 - \beta^2)^2 + (2\zeta_t \alpha\beta)^2}{\left[(\alpha^2 - \beta^2)(1 - \beta^2) - \alpha^2\beta^2\mu\right]^2 + (2\zeta_t\alpha\beta)^2(1 - \beta^2 - \beta^2\mu)^2}} \tag{4.26}$$

where:

$U_{st} = \frac{P_0}{K}$ represents the static displacement when the maximum load value (P_0) be applied to the structure and

U_0 : The maximum dynamic response of the structure under harmonic load.

$\beta = \frac{\omega}{\omega_{st}}$ is the force frequency ratio

$\alpha = \frac{\omega_t}{\omega_{st}}$ is the frequency ration of TMD to structure.

$\zeta_t = \frac{c_t}{2m_t w_t}$: Damping ration of TMD

In Fig. 4.4, the amplification factor is plotted for an undamped SDOF system that is coupled with a TMD. The TMD constitutes 5% of the system's mass ($\mu = 0.05$) and is precisely tuned to match the system's natural frequency, thereby setting $\alpha = 1$. The amplification factor, R, is then computed across a range of damping ratios for the TMD, spanning from zero to infinity.

In a system without structural damping, the response amplitude reaches infinity at the two resonant frequencies of the combined structure-TMD setup. As the TMD's damping approaches infinity, the two masses effectively merge into a singular entity, transforming the system into a SDOF configuration with an augmented mass of $1.05M$. In this state, the amplitude at the resonant frequency also becomes infinite. Consequently, between these two extremes, there must exist an optimal value of ζ_t where the peak amplitude is minimized.

The second study explores the behavior of a SDOF system equipped with a TMD, subject to harmonic excitation matching its natural frequency. This leads to resonance in an uncontrolled setting. The SDOF system has a mass of 2000 kg, natural frequency of 2, and a damping ratio of 0.01. The TMD has 1% of the SDOF's mass and is tuned to SDOF's natural frequency. The external excitation is a sinusoidal acceleration lasting 2 seconds, while the displacement response of the structure is recorded for a total of 20 seconds. This allows for an initial transient state in the first 2 seconds, followed by a steady-state condition when the external excitation ceases.

Performance metrics of the TMD under varying damping ratios—0, 0.05, 0.1, and 0.15—are examined. Fig. 4.5 contrasts the displacement response of the mass under three different TMD damping ratios against uncontrolled responses.

To more clearly assess the TMD's efficacy in reducing peak displacement, Fig. 4.6 also presents the absolute values of the peak responses. The first scenario involves a TMD with zero damping. While this configuration is most effective at lowering peak response during the excitation phase, its performance declines sharply once the system reaches steady state. Here, the TMD inadvertently serves as an external exciter for the structure. This underscores the importance of incorporating damping into the TMD system. In the subsequent scenarios, where the damping ratio incrementally rises from 0.05 to 0.15, a trade-off emerges: increas-

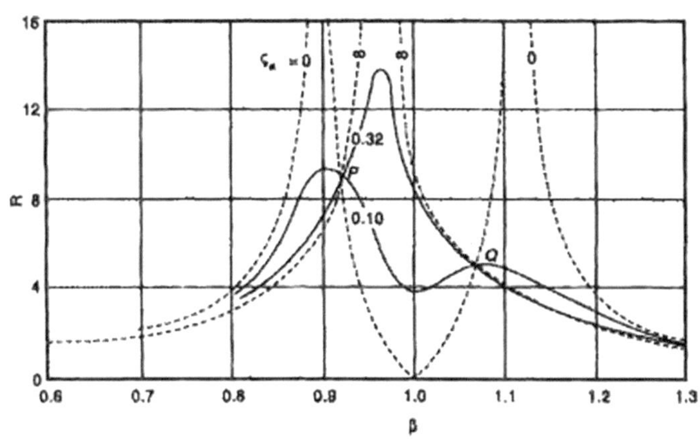

Fig. 4.4 Amplification factor as function of β ($\mu = 0.05$, $\alpha = 1$)

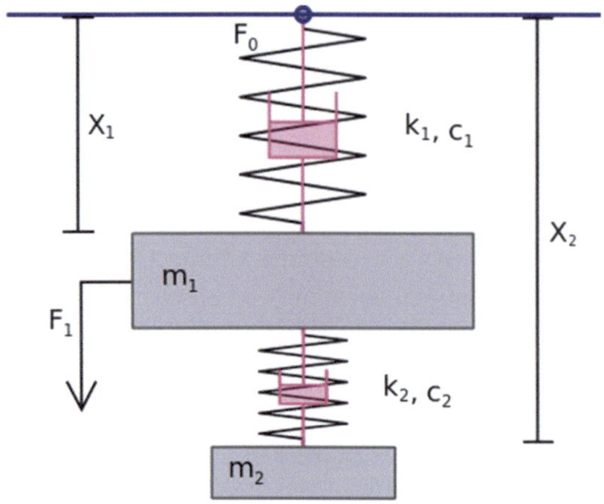

Fig. 4.5 SDOF system connected to a TMD

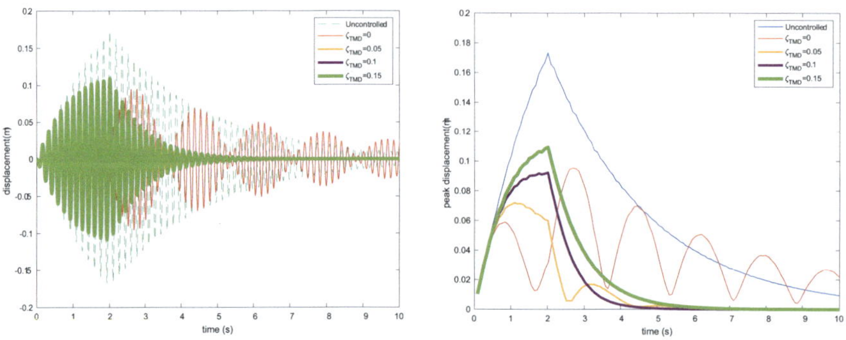

(a) Displacement response (b) Peak responses of the SDOF system

Fig. 4.6 Controlled and uncontrolled displacement response of a SDOF system to a harmonic excitation under different TMD damping ratios

ing damping improves performance during steady-state but reduce it during the transient phase. In other words, the TMD is more effective without external excitation but less so during the excitation period.

In summary, the selection of an optimal damping ratio plays an important role in the TMD's performance in steady state phase.

4.4 Evaluation of Controlled Structure

As earthquake and wind loads are stochastic in nature, somrtimes an analysis of the structure equipped by TMD system under properly scaled white noise excitation will shed more light on the effectiveness of TMD with regard to its application in wind and earthquake.

Assessing the performance of a structure equipped with TMDs is critical for ensuring both the effectiveness of the vibration control mechanism and the overall safety of the structure. Multiple evaluation indexes are commonly used to quantify this performance, allowing for a nuanced understanding of how well the TMD system is working. These indexes range from dynamic response metrics to safety and economic considerations. This section delves into the key evaluation indexes and the methodologies for applying them to the controlled structure.

Dynamic Response Metrics

a. Peak Displacement

The maximum displacement experienced by the structure under dynamic excitation serves as a vital indicator of the TMD's effectiveness [105].

b. RMS Displacement

The Root Mean Square (RMS) value of the structure's displacement is also an index of system behavior.

c. Peak Acceleration

Understanding the maximum acceleration experienced by the structure can provide insights into occupant comfort and structural integrity [102, 106].

d. Maximum Dynamic Stiffness

Optimizing for maximum dynamic stiffness ensures that the structure is robust against a wide range of dynamic loads [107].

e. Maximum Effective Damping of Combined Structure/TMD System

This criterion aims to maximize the damping effect brought about by the combination of the primary structure and the TMD [108].

f. Mixed Criterion Involving Frequency Tuning and Damping:

This multi-objective criterion involves tuning the TMD for minimum structural displacement while also optimizing for maximum effective damping [108].

g. Minimum Travel of Damper Mass Relative to Main Structure:

This criterion minimizes the relative motion between the TMD and the main structure, thus reducing wear and tear on the system [108].

h. Minimum Velocity of Main Structure

Minimizing the velocity of the primary structure can reduce the risks associated with rapid dynamic changes [106].

i. Minimum Stroke Forces

This criterion aims to minimize the stroke forces acting on the structure.

j. Response Spectrum

This metric shows the structure's maximum response to different frequency excitations, offering a holistic view of system performance.

k. Damping Ratio

This represents the degree of damping provided by the TMD, which is related to the reduction amount of number and magnitude of oscillations under external excitations.

Energy Metrics

a. Energy Dissipation

The amount of energy absorbed and dissipated by the TMD indicates its capability to control unwanted vibrations effectively.

b. Energy Transfer

This index assesses the efficiency with which energy is transferred from the structure to the TMD, crucial for optimal system performance.

Robustness and Reliability

a. Operational Range

Determining the range of frequencies over which the TMD effectively controls vibrations ensures its robustness.

b. Sensitivity Analysis

This involves examining how sensitive the TMD's performance is to variations in critical parameters, such as mass, damping, and stiffness.

Economic Criteria

a. Cost-Benefit Ratio

This metric takes into account both the initial investment in the TMD and the benefits accrued, such as increased structural lifespan or reduced maintenance costs.

b. Maintenance

Requirements Understanding the long-term maintenance needs of the TMD system aids in accurate economic planning.

Safety Metrics

a. Failure Modes

This involves identifying and assessing potential failure mechanisms of the TMD and the structure.

b. Regulatory Compliance

Ensuring the structure complies with local, national, and international building codes and standards is a non-negotiable aspect of any evaluation process.

Conclusion

Employing a multifaceted approach to evaluating a structure with TMDs enables a more thorough understanding of the system's performance. By considering these evaluation indexes, designers and engineers can more accurately gauge the effectiveness, safety, and economic viability of implementing TMDs in structural designs. By measuring the evaluation indexes through analytical methods, numerical simulations, and real-world tests, one can obtain a holistic view of the TMD's impact on the controlled structure.

Earthquakes and wind loads introduce additional complexities due to their stochastic nature. Subjecting a TMD-equipped structure to white noise excitation analysis can provide critical insights into its behavior under such unpredictable conditions. This stochastic analysis enhances the existing evaluation criteria, offering a more complete picture of the TMD's performance under the kinds of random, real-world external forces commonly encountered in wind and earthquake engineering.

4.5 Load Path Optimization

Establishing an effective load path is a critical phase in the design of a TMD. The primary objective is to ensure a robust and direct connection between the TMD supports and the building's lateral load-bearing system. A streamlined load path can sometimes be achieved by strategically integrating horizontal elements within the building's deck.

To bolster structural integrity, it is advisable to design all elements along the load path to withstand forces at least twice those predicted through analytical models. This provides an additional safety margin, ensuring stability against unforeseen impact forces.

Vertical Load Path Considerations

In addition to lateral forces, attention must be paid to designing an efficient vertical load path. Given that TMDs can be significantly heavy, localized stresses resulting from vertical earthquake-induced forces must be adequately accounted for. A structural engineer should meticulously analyze and design the associated load-bearing elements to minimize the risk posed by these vertical components.

Practical and Theoretical Examples 5

To solidify the presented techniques in this book, this chapter offers two chosen examples that showcase the practical and theoretical applications.

The first example delves into the realm of high-rise building design, with a focus on 3D modeling—closely aligning with real-world engineering challenges. It employs the optimum TMD parameters, as derived from the methodologies laid out in Chap. 4. Additionally, a hands-on approach to TMD modeling is presented.

The second example caters more to researchers, addressing the complex issue of multi-mode control in tall buildings through the use of a distributed Multiple Tuned Mass Damper (MTMD) system. It provides an in-depth walkthrough of using GA for solving intricate multi-objective optimization problems. This example is comprehensive, incorporating a variety of theories and methods discussed earlier in the book, including earthquake record selection and matching, dynamic analysis via Newmark's β method, the formulation of objective functions, the evaluation of the controlled system based on multiple criteria, and sensitivity analysis.

Studying these examples helps the reader to have a better understand the discussed techniques and also gain insights into how to apply them in both practical and research settings.

5.1 Single Mode Control of a High-Rise Building

5.1.1 Problem Description

In this example, the primary objective is to optimize the displacement responses of a 32-story high-rise building. The building has dimension of 30 m in length and 15 m in width, composed of six spans in the X-direction and three bays in the Y-direction, each measuring 5 m. Each story of the building has a height of 3 m, resulting in a total building height of 96 m.

H. Radmard Rahmani, *Seismic Control of Structures Using Tuned Mass Dampers*, Synthesis Lectures on Engineering, Science, and Technology,
https://doi.org/10.1007/978-3-031-71217-3_5

The building employs a steel structural system with distinct lateral load-bearing mechanisms: mega-bracing is used on the perimeter frames in the X-direction, while moment frames are implemented in the Y-direction. The goal is to enhance the building's structural performance using TMDs.

Accounting for the weight of the structure, which is approximately 10,774 tons, and incorporating an additional 30% to account for live loads, the calculated seismic mass for the structure stands at 12,070 tons.

5.1.2 Structural Model

The FE-model of structure is developed in RFEM software, as depicted in Fig. 5.1. Beam-column connections are modeled as hinged in the X-direction and fixed in the Y-direction. The column bases are assumed to have fixed restraints. Deck elements are simulated using shell elements, each with six degrees of freedom (6-DOF) at their nodes.

Fig. 5.1 Structural model of high-rise building

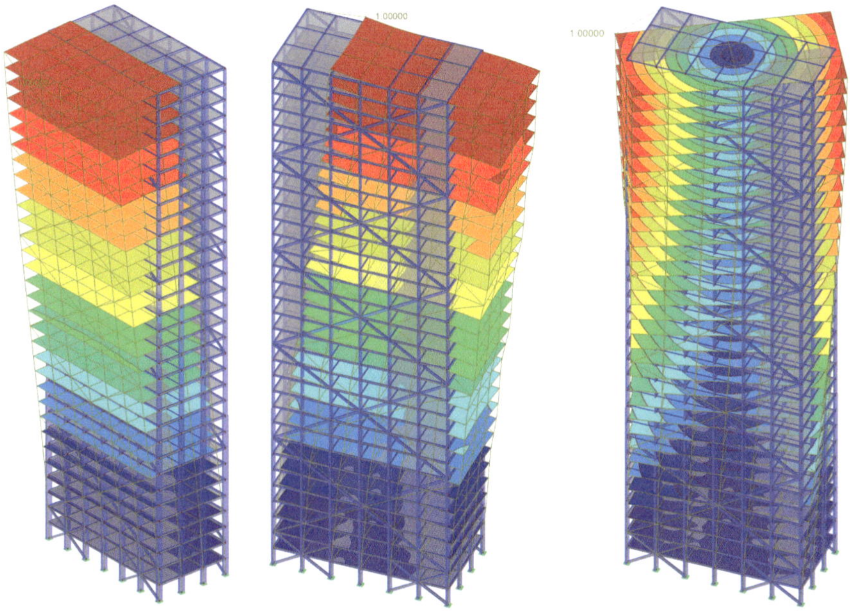

(a) Mode-1 (b) Mode-2 (c) Mode-3

Fig. 5.2 The first three structural modes of the structure **a** Mode-1: $f = 0.32$ Hz, **b** Mode-2: $f = 0.55$ Hz, **c** Mode-3: $f = 0.66$ Hz

For load considerations, a dead load of 6 kN/m^2 is distributed across the decks, while a live load of 3 kN/m^2 is also incorporated.

The structure's first three natural modes are illustrated in Fig. 5.2. The natural frequencies for these modes are 0.32 Hz, 0.55 Hz, and 0.66 Hz, respectively. The first mode corresponds to lateral displacements in the Y-direction, indicating relatively low stiffness in this axis. The second mode is associated with lateral displacements in the X-direction, while the third mode represents rotational movement. For this case study, the Northridge earthquake acceleration record serves as the external excitation in the Y-direction, as shown in Fig. 5.3.

Key modal properties of the structure for these initial three modes are compiled in Table 5.1.

The building's uncontrolled roof-level displacement in the Y-direction is captured in Fig. 5.4 The peak uncontrolled displacement response measures at 280 mm, followed by a secondary peak at 212 mm. The aim is to enhance the structural responses using a single TMD.

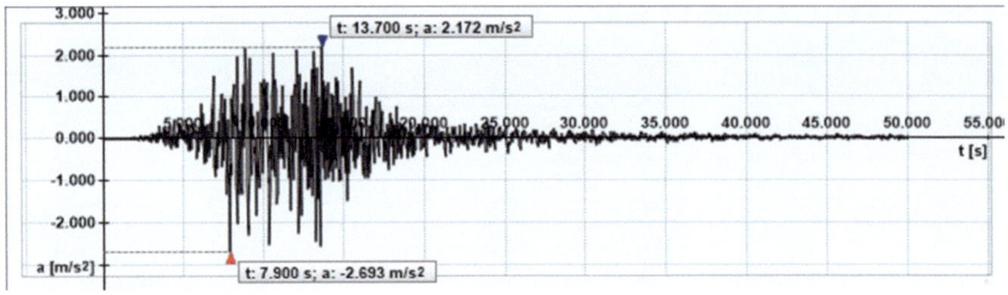

Fig. 5.3 Northridge earthquake acceleration record

Table 5.1 Modal parameters

Mode	Frequency (Hz)	Modal mass (t)
1	0.32	4183
2	0.55	4003
3	0.66	1885

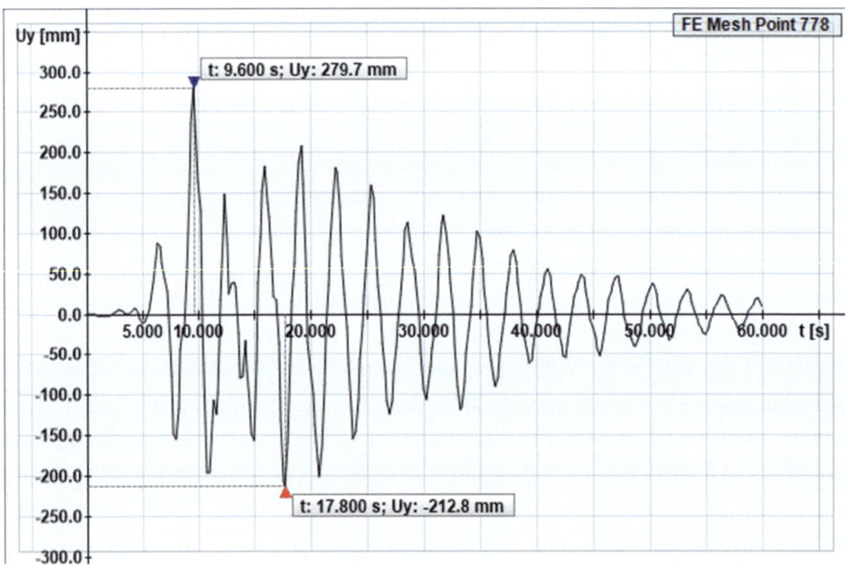

Fig. 5.4 The uncontrolled displacement response at the roof level of the building in Y-direction

5.1.3 TMD Design Considerations

The TMD is considered to be placed at the centroid of the mass on the building's top floor. Given the structure's first mode shape, the roof level exhibits the maximum modal displacement, making it an ideal location for the application of control forces. The TMD's mass is set at 236 tons, constituting approximately 2% of the structure's active seismic mass.

By taking into account this mass, and the frequency of the first mode—the target mode and by utilizing the tuning factors outlined in the TMD design section, the optimal parameters for the TMD are determined as follow:

$$\mu = 0.02$$

$$f_1 = 0.32\ Hz$$

$$\omega = 2\pi(0.32) = 2.02\ rad/s$$

$$\alpha_{opt} = \frac{\sqrt{1-\frac{\mu}{2}}}{1+\mu} = 0.975$$

$$\zeta_{opt} = \sqrt{\frac{\mu(1-\frac{\mu}{4})}{4(1+\mu)(1-\frac{\mu}{2})}} = 0.07$$

Therefore:

$$m_t = 0.02 \times 12070 = 241\ t$$

$$k_t = m_t \alpha_{opr}^2 \omega_{st}^2 = 241 \times 0.975^2 \times 2.02^2 = 940.3\ kN/m$$

$$c_t = 2\zeta_{opt} m_t \sqrt{\frac{k_t}{m_t}} = 2(0.07) \times 241 \times \sqrt{\frac{940.3}{241}} = 66.88\ kN.s/m$$

The TMD is modeled by allocating a mass to a node, which is then linked to the structure's centroid of mass at the roof level through a dashpot.

5.1.4 Controlled Responses

The controlled response of the structure is depicted in the Fig. 5.5. As illustrated, the peak displacement response at the building's top level has been reduced to 249 mm, reflecting a 10% improvement compared to the uncontrolled state. Notably, the second peak of the displacement response has been reduced to 158 mm, representing a 25% reduction.

Moreover, the diminishing magnitude of the peaks indicates a significant enhancement in the structure's damping capabilities. It's worth noting that TMDs may not be as effective at

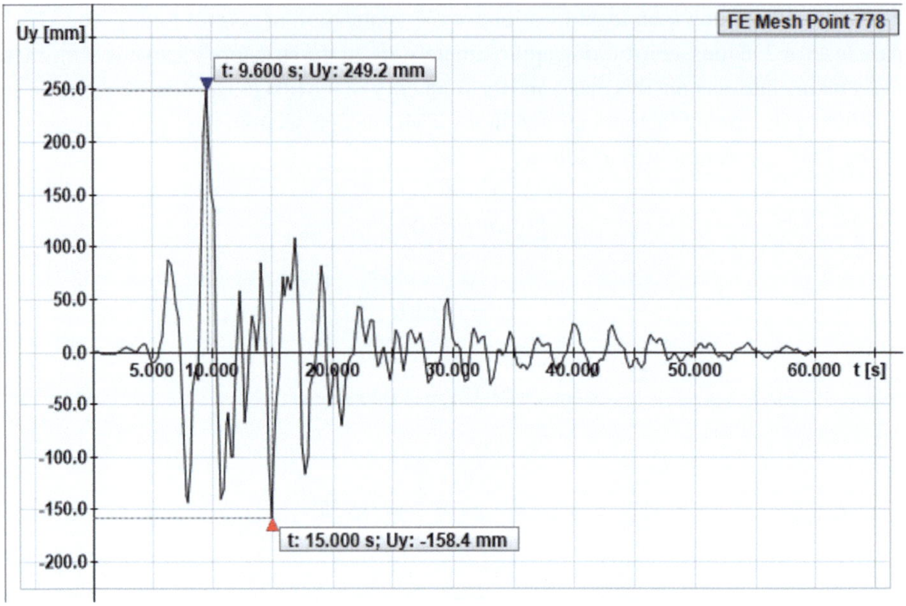

Fig. 5.5 The Controlled displacement response at the roof level of the building in Y-direction

mitigating the initial peaks; they require a few oscillations to achieve optimal performance by oscillating out of phase with the structure.

5.2 Multi-Mode Control of a Tall Building

5.2.1 Introduction

In this example, we aim to integrate the methodologies discussed in earlier chapters to address a more complex control issue: vibration control of a 76-story benchmark building under the influence of seven specifically selected and scaled earthquake excitations. A key distinction in this example is the heightened involvement of multiple structural modes in the building's overall response due to its considerable height. This leads us to explore whether multi-modal control could serve as an optimal strategy in such scenarios. Furthermore, we aim to determine the most effective locations for the TMDs.

Given the intricate nature of multi-modal control and the multitude of parameters involved, a try-and-error approach would prove inefficient for optimizing system performance. To navigate this complexity, we employ Non-dominated Sorting Genetic Algorithm II (NSGA II) [94], as a robust tool for multi-objective optimization problems. Objectives are pivotal criteria in any optimization scenario; they come into play when contrasting one solution against another to establish a preference. A central challenge lies in converting the structural engineering problem into a quantifiable mathematical optimization problem.

The intent of this section is to provide comprehensive answers to these questions while outlining the process involved. The results presented here are drawn from a study conducted by Rahmani et al. [96], which focuses on the seismic control of tall buildings via an enhanced version of NSGA II.

5.2.2 Overview

5.2.3 Problem Description

The overarching objective of this section is to enhance the seismic performance of a high-rise building through the optimized deployment of one or more TMDs. Specifically, we seek to determine the optimal parameters for the control system, which include the number of TMDs, their strategic placement within the structure, and the fine-tuning of their operational settings.

The subject of our study is a 76-story, 306-meter tall concrete edifice, based on a benchmark model developed by Yang et al. [95]. The structure has a total mass of 153,000 tonnes and occupies a volume of 510,000 cubic meters, resulting in a mass density of approximately 300 kg/m^3. The building's slender design is characterized by a height-to-width ratio of 7.3 (calculated as 306.1/42), making it particularly sensitive to wind forces for which it has been primarily designed to withstand.

In terms of mathematical modeling, the building is represented as a vertical cantilever beam. A finite element model is constructed by treating each segment between two consecutive floors as a classical beam element with uniform thickness. This configuration results in a system with 76 translational and 76 rotational degrees of freedom (DOFs). Subsequently, all rotational DOFs are eliminated through static condensation, leaving us with 76 translational DOFs that correspond to lateral displacements at each floor level.

The building exhibits its first five natural frequencies at 0.16 Hz, 1.992 Hz, 3.79 Hz, and 6.35 Hz, as illustrated in Fig. 5.6. To quantify the damping characteristics, a 76x76 proportional damping matrix is calculated using Rayleigh's method, assuming a 1% damping ratio for these first five modes.

By comprehensively understanding these parameters and characteristics, we aim to deploy TMDs in the most effective manner to mitigate seismic responses, leveraging optimization techniques for the best possible outcomes.

5.2.4 Mathematical Model of Building

The optimization process necessitates the assessment of the structure under varying sets of TMD parameters to ascertain its seismic responses. These responses serve as the basis

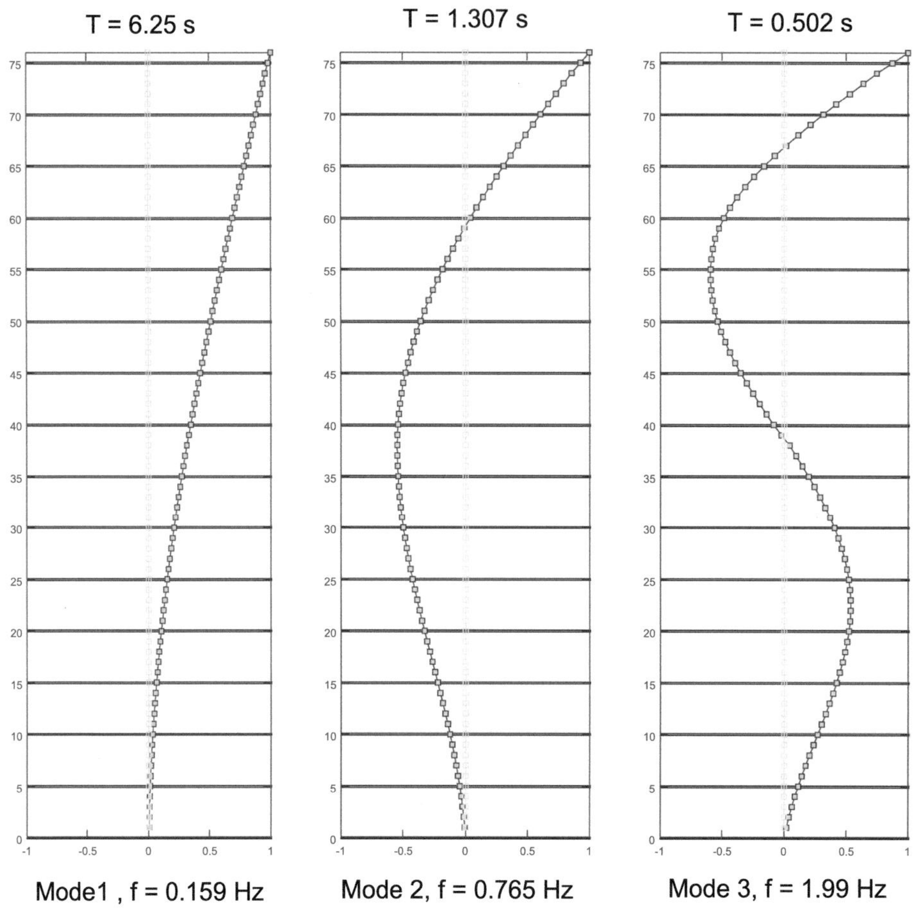

Fig. 5.6 Five mode shapes of the 76-story building

for calculating the objective functions that guide the GA in its iterative search for optimal solutions. Hence, a mathematical model of the building's dynamics is indispensable for this endeavor.

As previously discussed, the governing equation of motion for the high-rise building subjected to earthquake excitation can be represented as follows:

$$[M]\{\ddot{U}\} + [C]\{\dot{U}\} + [K]\{\dot{U}\} = \{P_t\} \tag{5.1}$$

where M, K and C represent mass, stiffness and damping matrices of the structure and the TMDs:

$$[M] = [M_{st}] + [M_t] \tag{5.2}$$

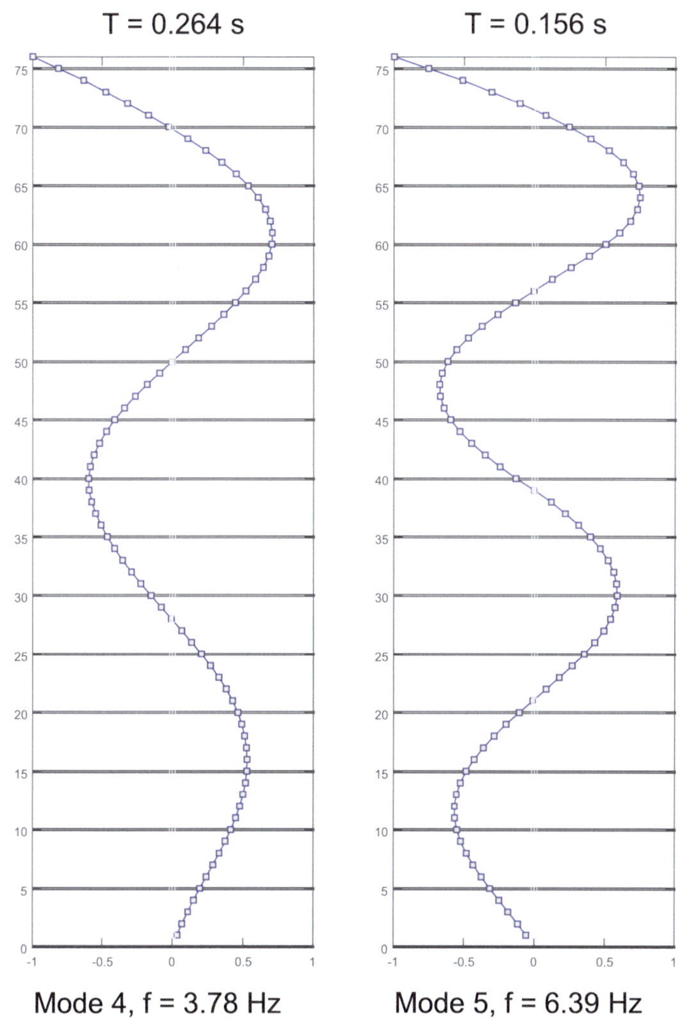

Fig. 5.6 (continued)

$$[C] = [C_{st}] + [C_t] \tag{5.3}$$

$$[K] = [K_{st}] + [K_t] \tag{5.4}$$

Indexes st and t indicate the DOFs of the building and the TMDs, respectively.

The external load vector, P_t in Eq. 5.1 comprises inertial forces due to ground acceleration as follows:

$$\{P_t\} = -\ddot{u}_g\, [M]\, \{1_t\} \tag{5.5}$$

where $\{1_t\}_{(N+n)\times 1} = [1\ 1\ \cdots\ 1]^T$ and the term $\ddot{u}_g$ represents the ground accelerations.

The structural responses, including displacement, velocity, and acceleration matrices are defined by:

$$\{U\} = \{u_{st1}, u_{st2}, \ldots, u_{stN}, u_{t1}, u_{t2}, \ldots, u_{tn}\} \tag{5.6}$$

$$\{V\} = \{v_{st1}, v_{st2}, \ldots, v_{stN}, v_{t1}, v_{t2}, \ldots, v_{tn}\} \tag{5.7}$$

$$\{A\} = \{a_{st1}, a_{st2}, \ldots, a_{stN}, a_{t1}, a_{t2}, \ldots, a_{tn}\} \tag{5.8}$$

In these equations, N and n represent the number of DOFs for the building and the TMDs respectively. Therefore, the dimensions of the M, K, and C matrices are $(N+n)\times(N+n)$.

The design parameters of a TMD include its damping, tuning frequency, and mass. Generally, the ratios of these parameters to the corresponding values of the building are considered as design parameters:

$$m_0 = \frac{m_t}{m_{st}},\ \beta = \frac{v_t}{v_{st}},\ \psi = \frac{c_t}{c_{st}} \tag{5.9}$$

Where m_0, β, ψ refer to mass, frequency, and damping ratios.

In order to solve the equations of the motion, Newmark's β method is utilized. Average acceleration method is considered by setting $\beta = \frac{1}{2}$ and $\beta = \frac{1}{4}$ in the relevant formulation [109].

5.2.5 Optimization Variables

The optimization variables are the values that we look for their optimums. The GA will change this variables and watches the objectives to determine how good are they selected. For multiple distributed tuned mass dampers system the following parameters are considered as optimization variables:

1. Number of TMDs
2. Position of the TMDs→ story number
3. TMDs' properties→ m_0, β, ψ

These parameters are schematically shown in Fig. 5.7.

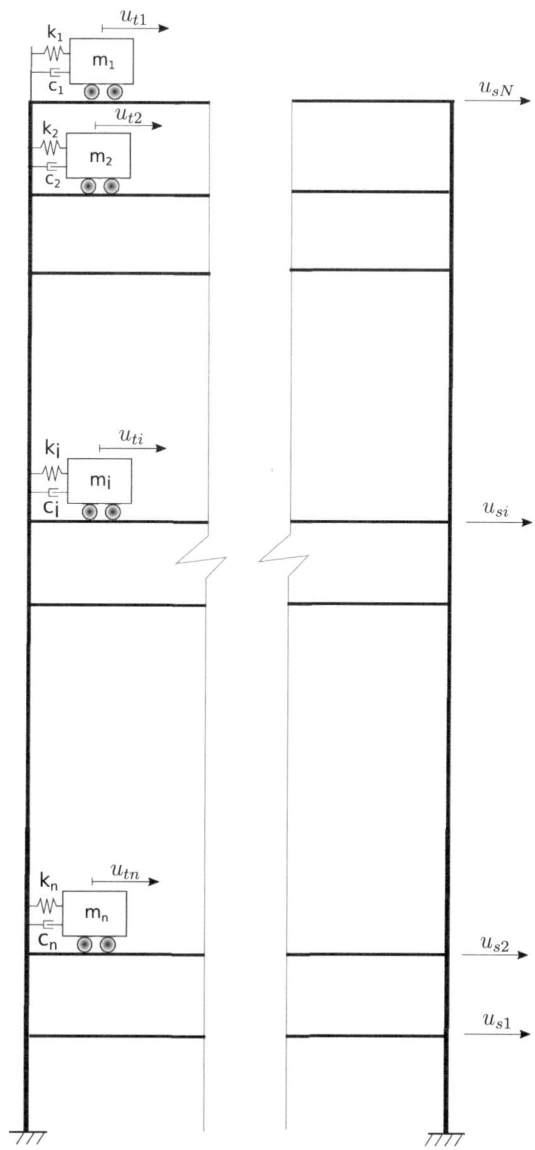

Fig. 5.7 Building equipped by TMDs

5.2.6 Genetic Algorithm for TMD Optimization

As previously emphasized, the design of TMDs revolves around identifying the optimal set of parameters that will maximize system performance for a specific structural issue. This quest for optimization generally aims to improve structural responses, such as displacement, velocity, and acceleration. Given the multidimensional nature of these objectives, we are inherently dealing with a complex optimization problem that is not easily solved through traditional methods.

We will utilize the GA, which has proven successful in addressing a wide range of complicated engineering challenges. In the ensuing section, we will recast the TMD design task into a mathematical optimization problem that can be effectively tackled using GA.

5.2.7 Encoding

In binary coding paradigm, each variable is represented as a binary string comprised of zeros and ones. The bit-length allocated for each variable is dictated by its potential range of maximum and minimum values. Ultimately, the resulting fixed-length binary string encapsulates the full spectrum of parameter values relevant to the problem at hand. In GA terminology, these binary segments reflecting TMD parameters are known as "genes." When aggregated, these genes form a "chromosome," which encapsulates all the requisite information for a given solution.

We will use binary coding for coding TMD parameters into a binary string with a constant number of genes, as is shown in Fig. 5.8. The parameters of a TMD is defined as follows:

$$m_0 = \frac{m_t}{m_{st}},\ \beta = \frac{\nu_t}{\nu_{st}},\ \xi = \frac{c_t}{c_{st}} \tag{5.10}$$

where the t and st indexes correspond to the TMD and the structure respectively.

Chromosomes are assembled by combinig all the genes corresponding to each solution. Consequently, each chromosome encapsulates the encoded data for all TMDs present in the building. Under this arrangement, the location of each TMD is represented by the corresponding gene set's position within the chromosome. A schematic depiction of these chromosomes is illustrated in Fig. 5.9.

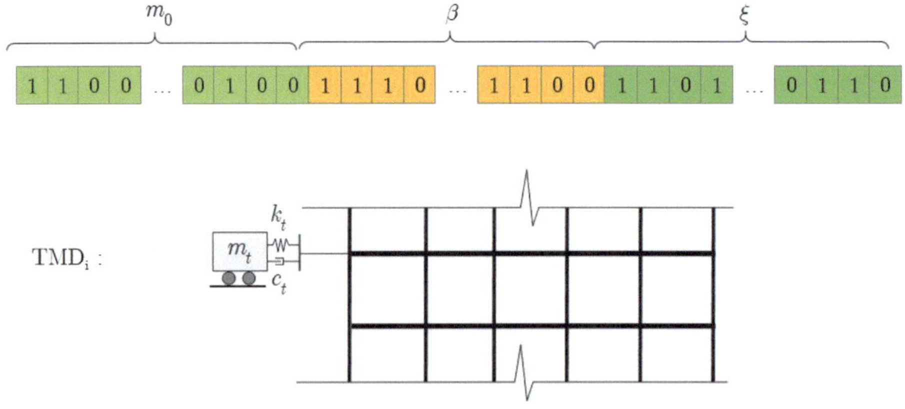

Fig. 5.8 Binary coding of a single TMD's parameters

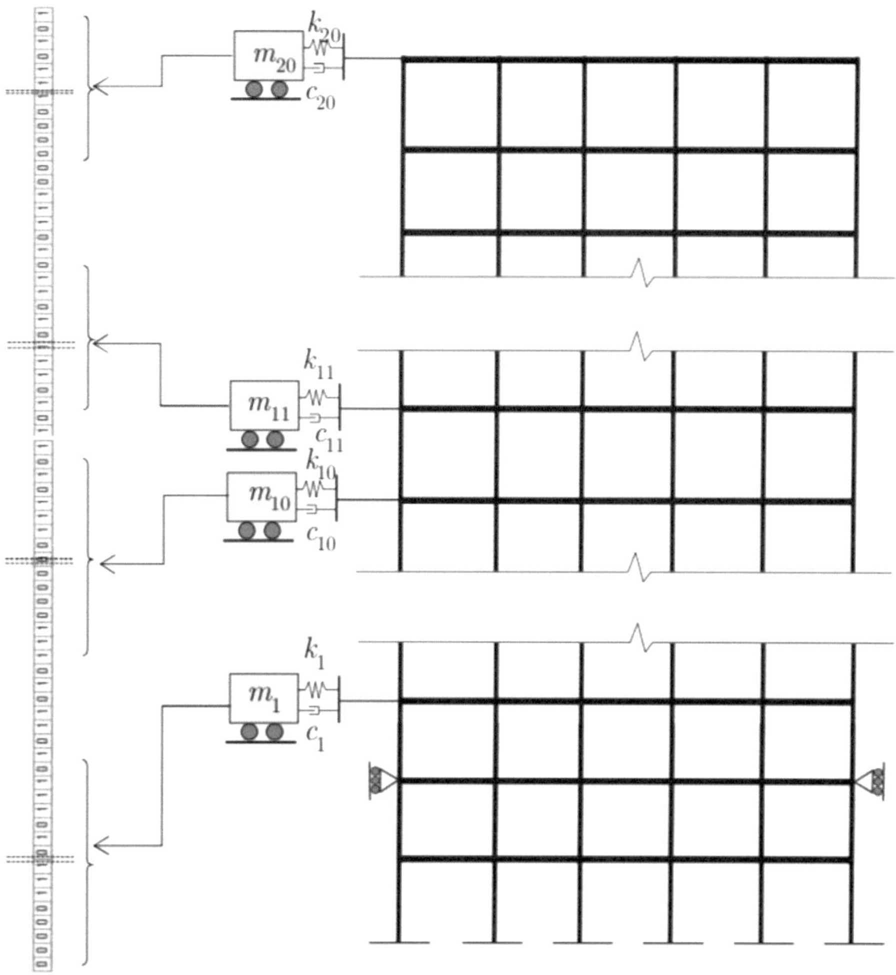

Fig. 5.9 A schematic chromosome representation

To proceed with this approach, it is crucial to establish a domain for each variable under consideration. The specific domains used in this study are outlined in Table 5.2. For the mass ratio, two constraints are considered: the maximum allowable mass ratio for individual TMDs, as well as a cumulative mass ratio limit $m_t = \sum_{i=1}^{n} m_{0i}$, capped at 3%. This flexibility permits the GA algorithm to allocate the available mass either to a single TMD or distribute it across multiple TMDs at varying heights within the building. Likewise, constraints are placed on the damping and frequency ratios of the TMDs to ensure they fall within practical limits.

Table 5.2 Parameters variation domain for TMDs

Parameters	Min. value	Max. value	Number of gens
m_t	–	3%	–
m_0	0.2%	3%	8
β	0.8	1.3	6
ψ	5	40	9

As detailed in Table 5.2, the binary representation for each TMD parameter is specified. In each 23-bit binary string, the first 8 bits are allocated for m_0, the subsequent 6 bits for β, and the remaining 9 bits for ψ. A notable challenge with binary encoding in this context is the representation of decimal ratios, as binary strings inherently yield integer values. To navigate this, a mapping function can be used to correlate each integer within the binary range to its corresponding decimal value within the predefined domain. For instance, a m_0 value of 1.1%, which lies within the specified range of [0.2, 1.3], could be encoded into an 8-bit binary string through this mapping function as follow:

$$n = round(\frac{1.1 - 0.2}{(1.3 - 0.2)} \times (2^8 - 1)) = 209$$

$$b_n = de2bi(209, 8) = \{00001011\}$$

Utilizing the same encoding scheme for the remaining two TMD parameters, a 23-bit binary string is generated to encapsulate the values for a single TMD. In the study, each floor of the 76-story building is allocated its own 23-bit string, culminating in a chromosome length of 1748 bits. The presence and location of TMDs within the building are inferred from the corresponding positions of TMD parameters within this chromosome; a zero value for a floor's TMD parameters indicates the absence of a TMD on that level.

In terms of encoding, it's advantageous to work with dimensionless variables. In the context of our problem, this is achieved by normalizing each TMD parameter relative to its corresponding structural value. Such normalization markedly enhances the efficiency of methods like GA, which are agnostic to the dimensions of the variables yet need to combine these parameters to generate offspring.

💡 When encoding parameters that can assume negative values, a dual-stage encoding technique can prove useful. Initially, a mathematical function is employed to map the parameter's range onto a strictly positive interval. The resulting positive values are then encoded in binary format. During decoding, the binary values are first translated back into their positive form before the inverse function is applied to recover the original parameter values.

5.2.8 Genetic Operators

a. Crossover function

As it is discussed in Sect. 3.12 the crossover function creates new solutions based on the existing solutions by exchange some gens on two chromosomes related to the solutions. In this example a two variation crossover function which is developed by Rahmani et al. [96] is utilized. As it is shown in Fig. 5.10, in the first variant, the algorithm, exchanges the parameters in one or multiple TMDs in a parent with the corresponding values in another parent. In the second variation, the algorithm changes a TMD in a parent with a TMD in

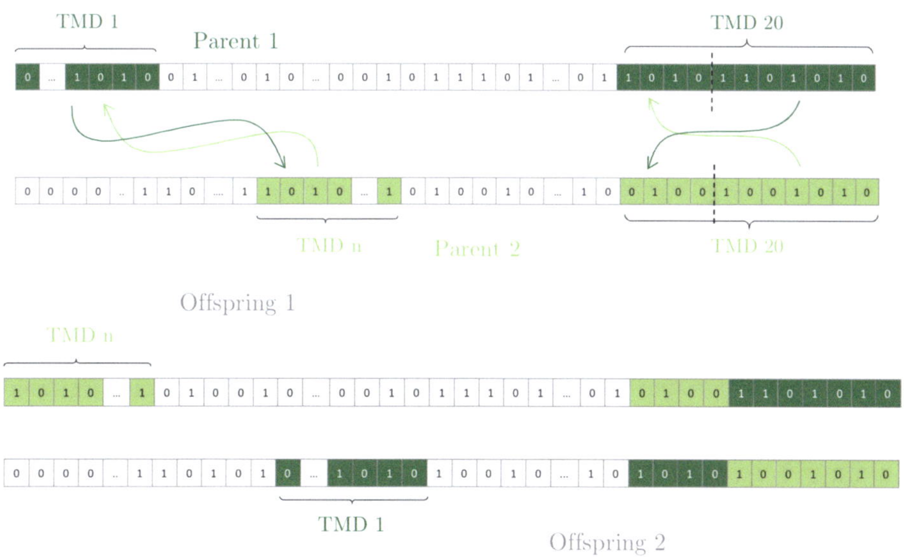

Fig. 5.10 Crossover types

another parent. The algorithm related to the two-variant crossover function is presented in Appendix A and A.2.

b. Mutation function

The Mutation Function generates a new solution by altering specific genes within an existing chromosome. Although conventional mutation functions flip bits in a binary-encoded chromosome, this approach can lead to impractical solutions in our context, such as TMDs with negligible mass or stiffness. To solve this, we employ a two-variant mutation function, as recommended by Rahmani et al. [96] . This specialized function acts upon two sets of genes: (1) those related to TMD properties, and (2) those specifying the TMDs' locations within the building (refer to Appendix A and A.3).

5.2.9 Repair Function

The Repair Function rectifies unfeasible solutions to ensure they comply with established constraints. An "unfeasible solution" in our scenario could include TMD properties that fall outside acceptable limits. If post-mutation chromosomes yield children with unrealistic mass values, for instance, the Repair Function is invoked. Detailed in 3.12.7, this function calculates the minimum distance between the unfeasible and feasible sets of TMD properties (m_0, β, ψ) and adjusts the chromosome accordingly.

5.2.10 Fitness Function

Once chromosomes have been selected and operated upon, resulting in an array of solutions with varying TMD configurations, it becomes essential to assess the quality of these solutions. In complex engineering endeavors like TMD optimization, we often confront multi-objective optimization challenges that require evaluating solutions against multiple criteria. The Fitness Function in GA serves to quantify the efficacy of each solution, providing a basis for subsequent comparisons.

From a structural engineering perspective, the three primary performance indicators for assessing the TMD system relate to the building's displacement, velocity, and acceleration responses, both with and without control measures. Accordingly, the following objective functions are commonly utilized in relevant studies:

$$J_1 = max.\left(\frac{u_i^C}{u_i^{UC}}\right)_{i=1,\ldots,N} \tag{5.11}$$

$$J_2 = max.\left(\frac{v_i^C}{v_i^{UC}}\right)_{i=1,\ldots,N} \tag{5.12}$$

$$J_3 = max.\left(\frac{a_i^C}{a_i^{UC}}\right)_{i=1,\ldots,N} \tag{5.13}$$

where:

i: Story number

N: Number of stories in the tall building

u_i^{UC}: Uncontrolled displacement response of story i

u_i^C: Controlled displacement response of story i

v_i^{UC}: Uncontrolled velocity response of story i

v_i^C: Controlled velocity response of story i

a_i^{UC}: Uncontrolled acceleration response of story i

a_i^C: Controlled acceleration response of story i

5.2.11 Overview the Optimization Using Genetic Algorithm Process

In this section, we outline the optimization process to elucidate how GA can facilitate the optimal positioning and parameterization of TMDs in structural engineering challenges.

The process begins with the establishment of a simulation module that the GA can interface with. This module is built upon the state-space mathematical model of the structure, assuming time-invariant structural parameters such as mass, stiffness, and damping matrices. The simulation takes TMD parameters and positions as inputs and produces time-series data on structural responses—specifically, displacements, velocities, and accelerations at the DOFs—as outputs.

Once the simulation module is in place, the GA initiates the optimization cycle by generating an initial population of random chromosomes, each encoding a unique set of TMD parameters and positions using binary code. For example, an initial population might consist of 100 such chromosomes. The GA then decodes each chromosome to extract the corresponding TMD parameters and positions, which are fed into the simulation module. This module updates the structural matrices based on the provided TMD configurations and conducts a numerical analysis, often using methods like the Newmark-β method, to calculate structural responses.

These calculated responses are subsequently evaluated by the Fitness Function, which quantifies the efficacy of controlled vs. uncontrolled structural behavior. Fitness scores are appended to each chromosome, which are then ranked; a lower rank signifies a more fit or 'better' solution in terms of structural control. The GA then performs selection, crossover, and mutation operations to generate a new set of chromosomes, favoring those with better fitness scores for continued evolution.

This new generation is again evaluated and ranked, and the least fit members are pruned to maintain a constant population size. The generation count is incremented, and the process iterates until specified stopping criteria are met. These criteria may include reaching a predetermined number of generations or observing an improvement between successive generations that falls below a certain threshold. Employing the NSGA-II algorithmic approach ensures convergence, as each new generation is at least as fit as the preceding one, if not more so.

5.2.12 Framework

Building on the methods and objectives previously discussed, a comprehensive framework is developed to optimize both the placement and characteristics of TMDs in tall buildings. The computational framework is implemented using Matlab and consists of the following input parameters:

1. **Structural Parameters:** These include the mass, stiffness, and damping matrices of the building.
2. **TMD Parameters:** Constraints on mass, damping, and stiffness, which should be defined based on specifications provided by vendors.
3. **Earthquake Acceleration Records:** Historical or simulated data as external excitation

The framework outputs the most optimal locations and specifications for the TMDs, along with a detailed performance evaluation of the system. The codes and algorithms are presented in appendix of the book.

The core algorithm within this framework operates through a main iterative loop, advancing the genetic generations. In the first cycle, the algorithm creates an initial population by generating random chromosomes, which correspond to random values of the specified variables. Subsequent iterations introduce new chromosomes, generated through genetic operations on existing members of the population. These chromosomes are then fed into the analysis module, which simulates how the structure, fitted with the proposed TMD system, responds to earthquake-induced vibrations.

Following this, a fitness module assesses each chromosome based on the structural responses, and passes the sorted population to an NSGA-II module. This latter module uses historical data on TMD configurations and their associated fitness values to construct Pareto fronts, which are then used to further refine the chromosome rankings. The algorithm's pseudo-code can be found in Appendix A- under Algorithm A.1.

Advanced programming techniques such as parallel computing can be employed to enhance the efficiency of the code and minimize computational time.

5.2.13 Ground Motion Selection

In this example, the ground motions are chosen and calibrated according to the intensity-based assessment procedures stipulated in the ASCE/SEI 07-10 standards [110]. For this example, we have selected seven earthquake acceleration records from the Pacific Earthquake Engineering Research (PEER) Center's NGA Strong Motion Database [8]. These records were then appropriately scaled to align with a pre-defined design response spectrum, as illustrated in Fig. 5.11. The seismic parameters under consideration, along with the adopted design response spectrum, are comprehensively detailed in Table 5.3.

The specifications of the non-scaled and scaled selected earthquakes are presented in Tables 5.4 and 5.5. The spectrum of the non-scaled and scaled excitations are shown in Fig. 5.12.

5.2.14 TMD Parameters

The permissible ranges for the TMD parameters m_0, β, and ψ are outlined in Table 5.6, reflecting constraints that are practically applicable. The cumulative mass ratio of the TMDs, denoted as $m_t = \sum_{i=1}^{n} m_{0i}$, is limited to 3%. This constraint aligns with the individual mass limit set for each TMD. This flexible arrangement allows the optimization algorithm the latitude to either allocate the entire permissible mass to a single TMD or to distribute it across multiple TMDs at varying elevations within the building.

5.2.15 Sensitivity Analysis of GA Parameters

Given that the parameters of the GA are intrinsically problem-specific [111], a meticulous sensitivity analysis was conducted to identify their optimal values for our particular study. This involved adjusting the crossover and mutation probabilities iteratively and evaluating their impact on the objectives, using the El Centro earthquake excitation as the test case.

The subsequent results were organized through the application of the NSGA-II sorting algorithm, from which Pareto fronts were generated. These are presented in Table 5.8. The analysis revealed that the optimal probabilities for crossover and mutation stand at 0.7 and 0.2, respectively. Implementing these optimized parameters not only enhanced the quality of the solutions generated but also improved the overall performance efficiency of the algorithm (Table 5.7).

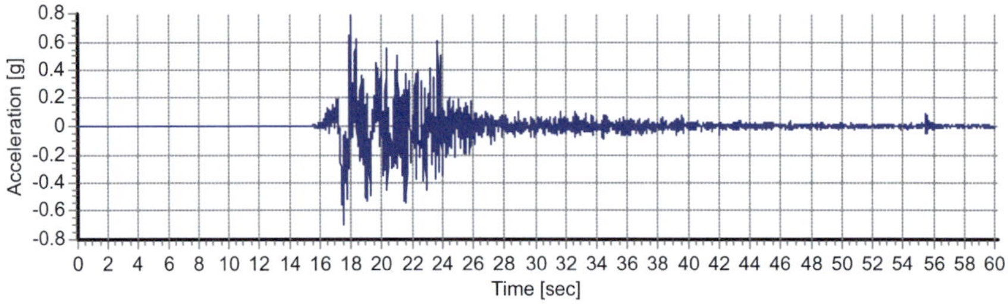

(a) Bam, 2003- Iran

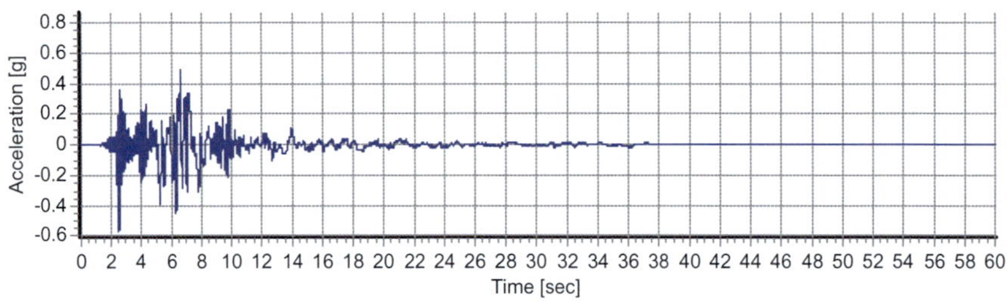

(b) Elcentro,1940- USA

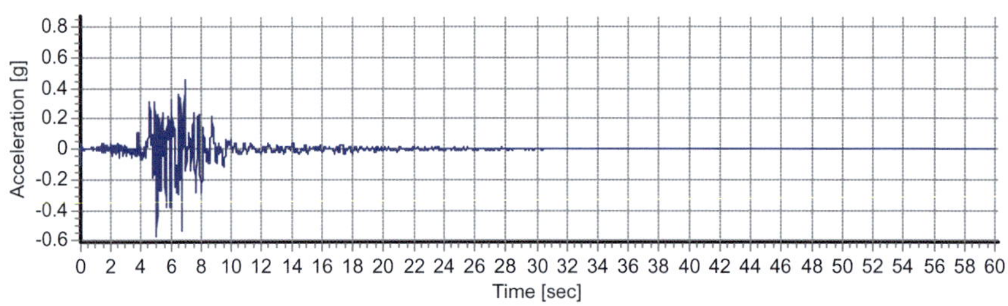

(c) Kobe, 1995- Japan

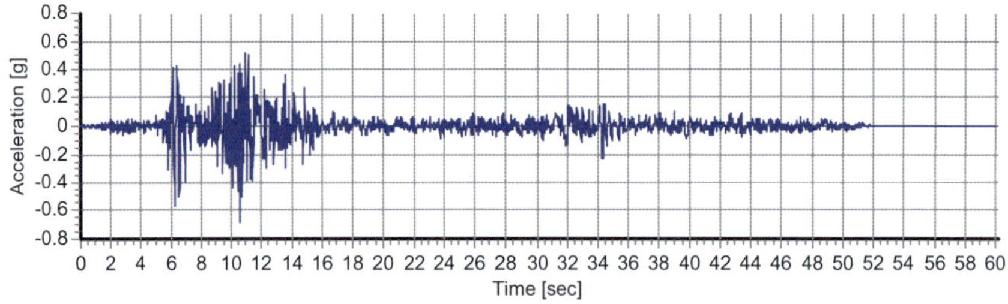

(d) Manjil, 2002, Iran

Fig. 5.11 Earthquake acceleration records obtained from the pacific earthquake engineering research (PEER) center [8]

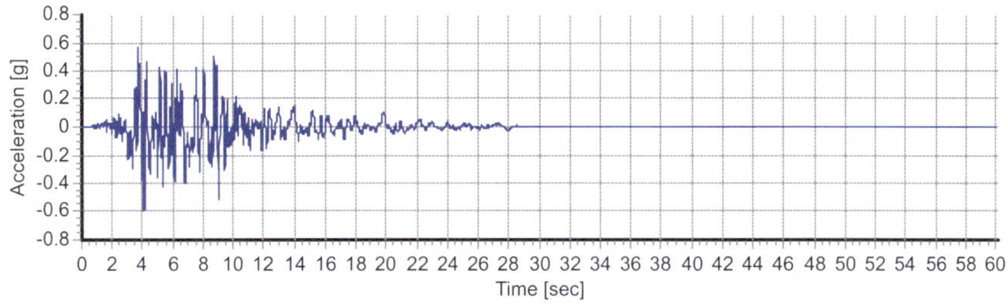

(e) Northridge, 1971- USA

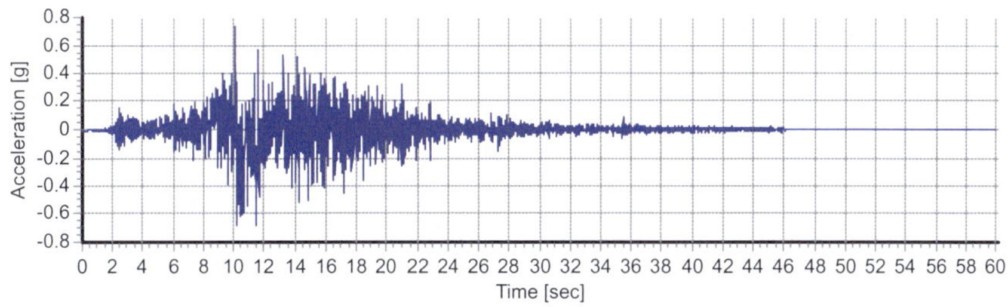

(f) Landers, 1992-USA

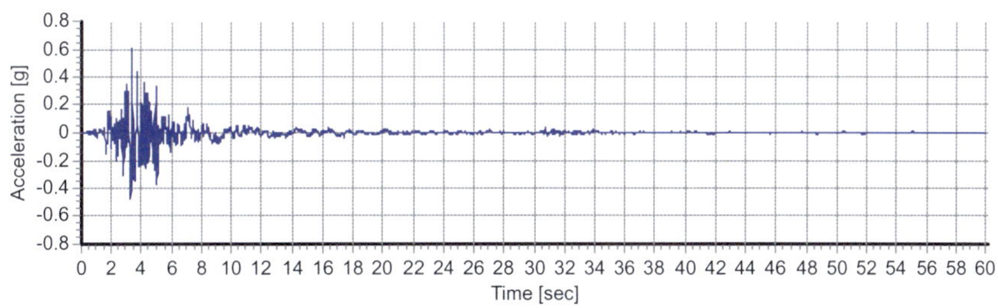

(g) SanFernando, 1994-USA

Fig. 5.11 (continued)

Table 5.3 Parameters of the design response spectrum

Site class	PGA	S_s	S_1	F_a	F_v	S_{MS}	S_{M1}	S_{DS}	S_{D1}
B	0.919	2.431 g	0.852 g	1	1	2.431 g	0.852 g	1.621 g	0.568 g

As a result, the parameters of the NSGA-II algorithm were considered as they are shown in Table 5.9. The sufficiency of 500 generations as the limit for the number of generations was then evaluated, as shown in Fig. 5.13.

Table 5.4 Original earthquakes' specifications (Dis. = Displacement (cm), Vel. = Velocity (m/s), Acc. = Acceleration (g))

Accelerogram	Max. acc.	Max. vel.	Max. dis.	Effective design acc.	Predominant period (sec)	Significant duration (sec)
1- Bam	0.80	124.12	33.94	0.69	0.20	8.00
2- Elcentro	0.44	67.01	27.89	0.30	0.06	11.46
3- Kobe	0.31	30.80	7.47	0.28	0.42	6.20
4- Manjil	0.51	42.45	14.87	0.47	0.16	28.66
5- Northridge	0.45	60.14	21.89	0.45	0.42	10.62
6-Landers	0.72	133.40	113.92	0.52	0.08	13.15
7- SanFernando	0.22	21.71	15.91	0.20	0.00	13.15

Table 5.5 Scaled earthquakes' specifications (Dis. = Displacement (cm), Vel. = Velocity (m/s), Acc. = Acceleration (g))

Accelerogram	Max. acc.	Max. vel.	Max. dis.	Effective design acc.	Predominant period (sec)	Significant duration (sec)
1- Bam	0.78	128.55	33.95	0.64	0.20	8.24
2- Elcentro	0.57	75.76	28.36	0.48	0.32	9.16
3- Kobe	0.56	38.83	15.68	0.56	0.36	4.16
4- Manjil	0.68	42.06	15.02	0.55	0.08	28.26
5- Northridge	0.59	64.57	22.16	0.58	0.40	10.40
6-Landers	0.73	141.02	113.78	0.56	0.08	12.92
7- SanFernando	0.60	23.71	15.88	0.61	0.10	6.03

Table 5.6 Parameter variation domain for TMDs

Parameter	Min. value	Max. value
m_t	–	3%
m_0	0.2%	3%
β	0.8	1.3
ψ	5	40

5.2.16 Optimal Arrangement and Parameters of TMDs: Results

Upon the successful preparation of the required inputs, including structural parameters and earthquake excitation records, the framework was employed to identify the optimal

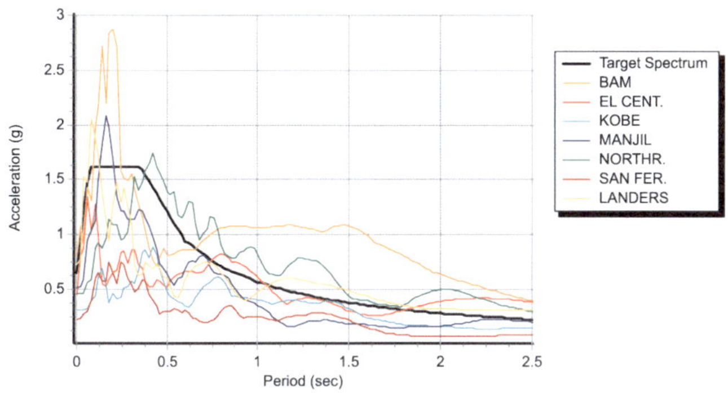

(a)

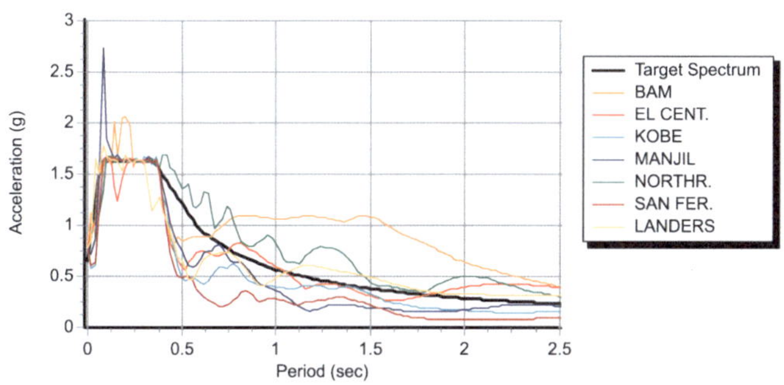

(b)

Fig. 5.12 Response spectrum for **a** unscaled and **b** scaled earthquake records

configuration and attributes of the TMDs (Figs. 5.15, 5.16, 5.17, 5.18 and 5.19). The results are delineated in Figs. 5.14 through 5.20.

Varied Excitation Patterns and TMD Configurations:

The analysis reveals that the optimal number of TMDs varies according to different excitations. Specifically, for the Bam and Manjil earthquake excitations, a maximum of three TMDs were deemed optimal. These were positioned at the 76th, 75th, and 64th levels of the building. The tuning frequencies for the top two TMDs closely matched the fundamental frequency of the structure, while the TMD at the 64th story was tuned to approximately 1.24 times the fundamental frequency.

Table 5.7 Crossover and Mutation variations

Variation no.	Crossover	Mutation	J1	J2	J3
1	0.6	0.1	0.845	0.917	0.941
2		0.2	0.860	0.926	0.950
3		0.3	0.855	0.924	0.948
4		0.4	0.839	0.915	0.942
5	**0.7**	0.1	0.862	0.925	0.946
6		**0.2**	**0.835**	**0.913**	**0.939**
7		0.3	0.861	0.925	0.947
8		0.4	0.846	0.919	0.944
9	0.8	0.1	0.869	0.929	0.950
10		0.2	0.876	0.932	0.952
11		0.3	0.842	0.915	0.941
12		0.4	0.852	0.922	0.947
13	0.9	0.1	0.871	0.930	0.951
14		0.2	0.866	0.927	0.947
15		0.3	0.874	0.932	0.953
16		0.4	0.839	0.913	0.939

Table 5.8 NSGA of Crossover and Mutation variations

Pareto front	Variation no.	Variation no.
1	**6**	–
2	16	–
3	4	11
4	1	–
5	8	–
6	5	12
7	3	7
8	2	14
9	9	–
10	13	–
11	10	15

Damping Ratios:

In terms of damping ratios, the TMDs on the upper two stories gravitated toward the maximum allowable value. However, the TMD located at the 64th level had a damping ratio of approximately 14.

Table 5.9 NSGA II algorithm initial parameters

Number of generations	Population size	Crossover probability	Mutation probability
500	100	0.7	0.2

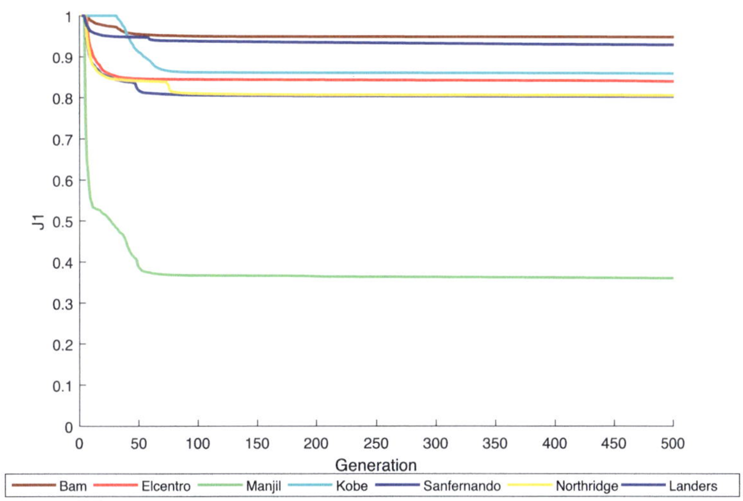

Fig. 5.13 Trend of objective J_1 during the generations

Performance of Control System:

A substantial improvement of around 65% is observed in the controlled displacement responses under the Manjil earthquake. In contrast, for the Bam earthquake, the objective J_1 registered a value of about 0.95, amounting to a modest 5% improvement in maximum displacements. A similar low efficacy was noted for the Landers earthquake, with a J_1 value of approximately 0.93, corresponding to a 7% improvement. Nevertheless, in both these cases, significant enhancements were made in damping further oscillations compared to uncontrolled states.

Other Earthquakes:

For the Landers, Northridge, and San Fernando earthquakes, the TMDs were placed optimally at the 76th and 74th levels. A majority of the allocated mass was devoted to the rooftop TMD. The optimum tuning frequencies were closely aligned with the fundamental frequency of the building, and the damping ratios ranged from 36.39 to 39.78.

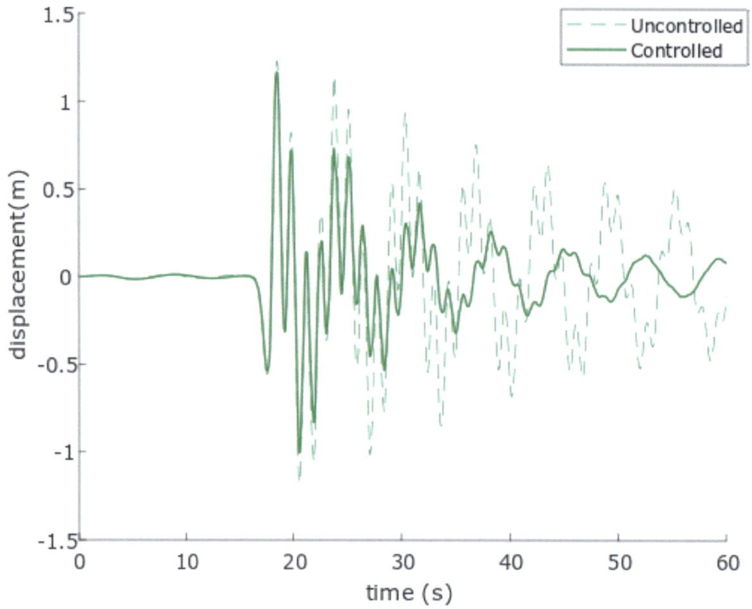

(a) Roof displacement response

Story	m_0	β	ψ
76	1,43	0.98	39,53
75	0.93	1.01	39,46
64	0,64	1.24	18.37

(b) TMDs specifications

J1	J2	J3
0,95	0,88	0,83

(c) Objective values

Fig. 5.14 Uncontrolled/controlled responses- Bam earthquake

Single TMD Systems:

For both the El Centro and Kobe earthquakes, the optimum configurations involved placing a solitary TMD system on the roof. In these instances, the entire permissible mass was allocated to the single TMD. The tuning frequencies for these TMDs hovered around 1.06 and 1.05, indicating a close alignment with the building's fundamental frequency.

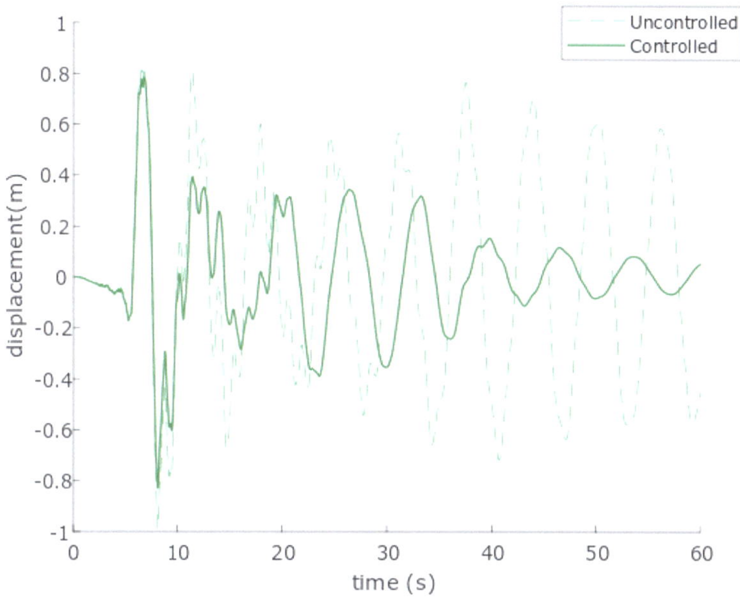

(a) Roof displacement response

Story	m_0	β	ψ
76	2,99	1,06	39,53

(b) TMDs specifications

J1	J2	J3
0,84	0,91	0,93

(c) Objective values

Fig. 5.15 Uncontrolled/controlled responses- Elcentro earthquake

5.2.17 Understanding the Optimal Placement of TMDs: A Modal Analysis

As detailed in the results section, the most advantageous locations for installing the TMDs consistently included the top two levels of the building for all earthquake excitations. Additionally, certain specific levels—namely the 74th and 64th floors—were also identified as optimal for some earthquake scenarios. To delve into the rationale behind these placements, we revisit the mode shapes of the building, as illustrated in Fig. 5.21, where stories experiencing maximum displacements are highlighted.

Modal Displacement Insights:

The analysis revealed that for the building's first three vibrational modes, the top three levels experienced the most significant displacements. This confirms the reasoning for positioning

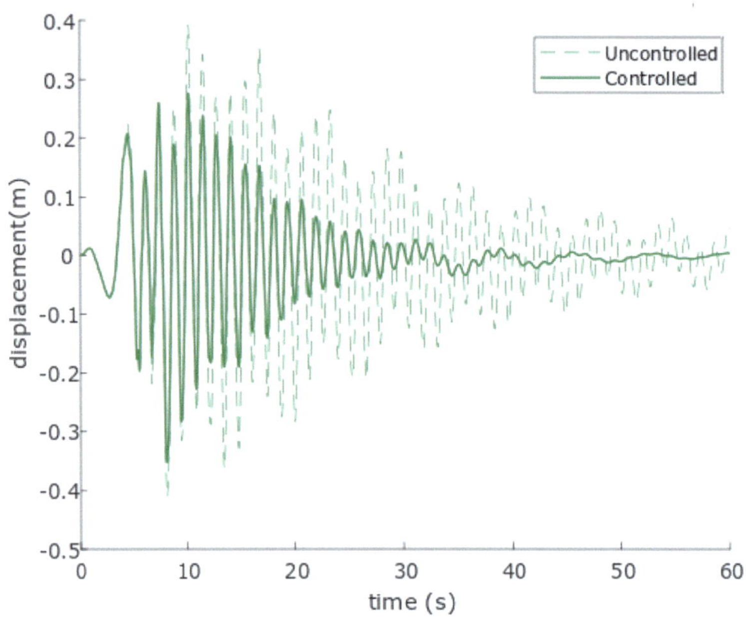

(a) Roof displacement response

Story	m_0	β	ψ
76	3,00	1,05	39,78

(b) TMDs specifications

J1	J2	J3
0,86	0,90	0,97

(c) Objective values

Fig. 5.16 Uncontrolled/controlled responses- Kobe earthquake

TMDs at these heights. In the case of the fourth vibrational mode, the most considerable displacements occurred at the rooftop, followed by the 75th and 61st levels. Similarly, the fifth vibrational mode saw peak displacements at the roof, the 75th, and the 64th levels.

These observations underscore the alignment between the stories experiencing maximum modal displacements and those chosen as the optimal locations for TMD installations. The findings effectively validate the strategic positioning of TMDs in the structure to maximize their efficacy in dampening vibrations and enhancing structural stability.

Exploring the Role of Excitation Properties in TMD Placement

While the optimal placement of TMDs on the top stories effectively mitigates modal displacements across all five modes—aligning with our optimization results—certain earth-

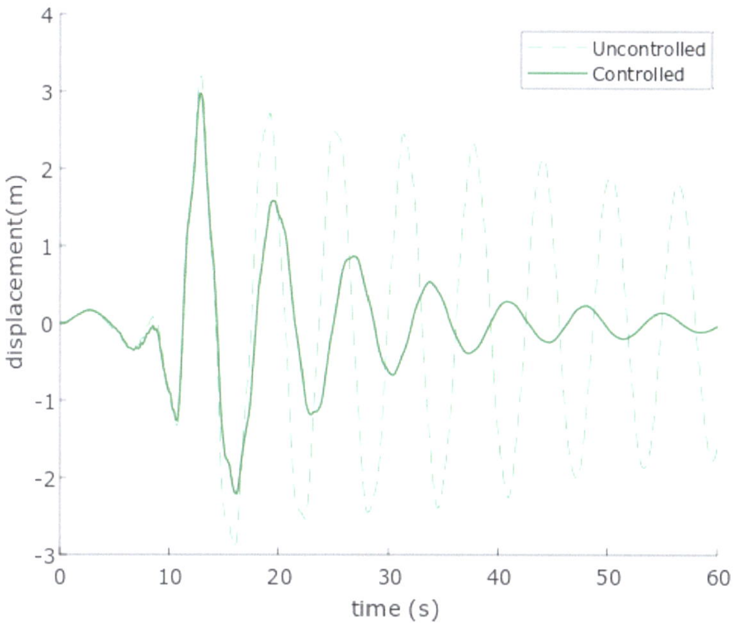

(a) Roof displacement response

Story	m_0	β	ψ
76	2,01	1,01	37,07
74	1,00	1,01	37,13

(b) TMDs specifications

J1	J2	J3
0,93	0,89	0,98

(c) Objective values

Fig. 5.17 Uncontrolled/controlled responses- Landers earthquake

quake scenarios required TMD placements at lower stories. This suggests that the ideal location for TMDs is not solely dependent on the building's structural characteristics but also on the specific properties of the earthquake excitation.

To further investigate this relationship, Fourier Transform analyses were carried out on the excitation records for each earthquake. These analyses, depicted in Fig. 5.22, specifically examined the amplitudes corresponding to the building's natural frequencies.

Intriguingly, for the Bam and Manjil earthquakes, the excitation amplitudes in the 4th and 5th modes exceeded those in the lower modes. Consequently, despite these higher modes having relatively lower mass participation factors, their contribution to the overall earthquake response is amplified due to these higher excitation amplitudes.

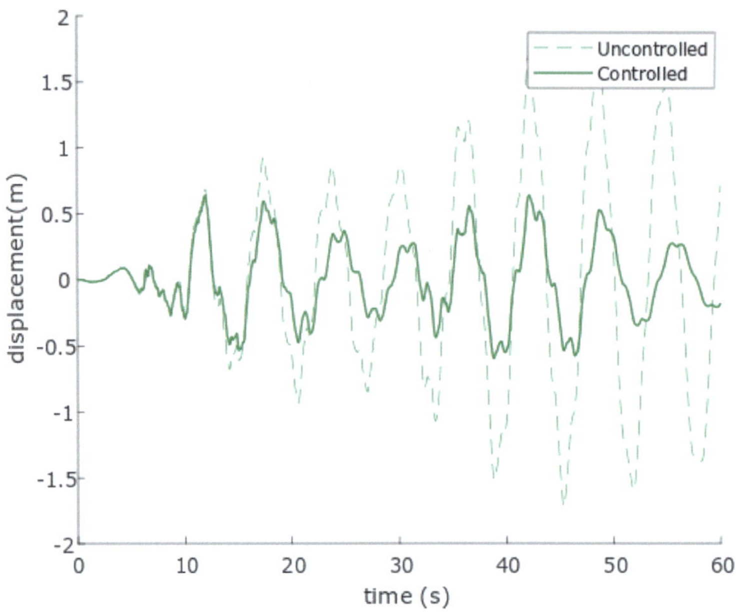

(a) Roof displacement response

Story	m_0	β	ψ
76	1,50	1,02	35,12
75	0.95	1,01	31,34
64	0,55	1.21	14,92

(b) TMDs specifications

J1	J2	J3
0,36	0,57	0,95

(c) Objective values

Fig. 5.18 Uncontrolled/controlled responses- Manjil earthquake

This nuanced finding illuminates the complexity of optimal TMD placement, emphasizing that both structural modal behavior and the unique frequency content of the seismic excitations must be taken into account for effective vibration control.

The Role of Modal and Seismic Factors in Building Response

To theoretically investigate the obtained results, the displacement response of the building under the ground motion is presented in Eq. 5.14 as sum of the modal nodal displacements:

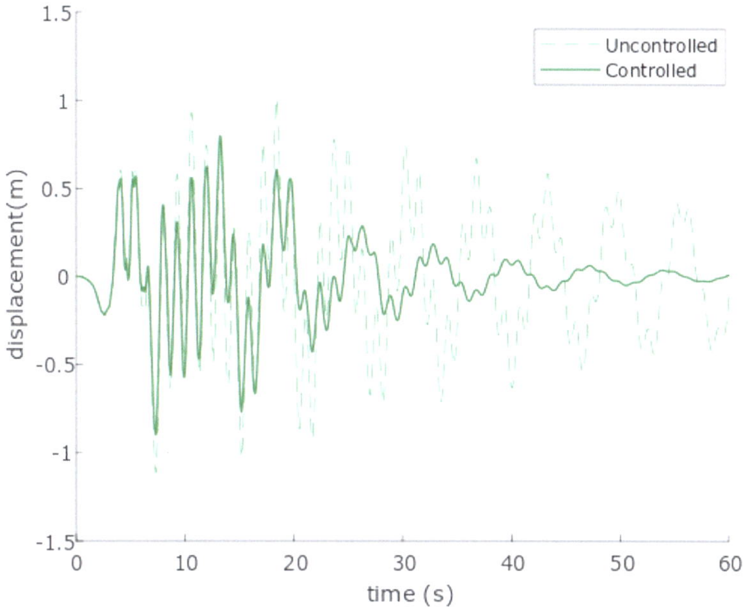

(a) Roof displacement response

Story	m_0	β	ψ
76	2,26	1,04	37,05
74	0,75	1,04	36,98

(b) TMDs specifications

J1	J2	J3
0.81	0.84	0.95

(c) Objective values

Fig. 5.19 Uncontrolled/controlled responses- Northridge earthquake

$$\mathbf{u}(t) = \sum_{n=1}^{N} \mathbf{u}_n(t) \tag{5.14}$$

where $\mathbf{u}_n$ represents the mode nth displacements. The contribution of the nth mode to the nodal displacement $\mathbf{u}$(t) is

$$\mathbf{u_n}(t) = \phi_n q_n(t) \tag{5.15}$$

where q_n refers to the modal coordinate which can be calculated from following equation:

$$\ddot{q_n} + 2\zeta_n \omega_n \dot{q}_n + \omega_n^2 q_n = -\Gamma_n \ddot{u}_g(t) \tag{5.16}$$

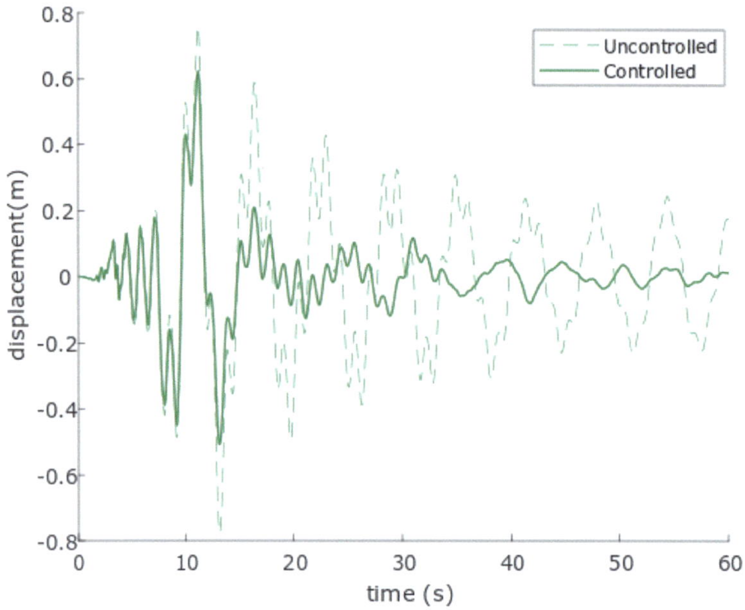

(a) Roof displacement response

Story	m_0	β	ψ
76	2,28	1,03	36,27
74	0,72	1,03	36,39

(b) TMDs specifications

J1	J2	J3
0.80	0.84	0.97

(c) Objective values

Fig. 5.20 Uncontrolled/controlled responses- SanFernando earthquake

Γ_n is the modal participation factor of the nth mode and is the degree to which the nth mode participates in the total response. The modal participation factor can be calculated based on the modal displacements and masses as follows:

$$\Gamma_n = \frac{\{\phi_n\}[M]\{1\}}{\{\phi_n\}^T[M]\{\phi_n\}} = \frac{\sum_{j=1}^{N} m_j \phi_{jn}}{\sum_{j=1}^{N} m_j \phi_{jn}^2} \tag{5.17}$$

Equation 5.16 outlines the behavior of a SDOF system, characterizing its frequency and damping in relation to the nth mode.

As demonstrated, nodal displacement $\mathbf{u}_n(t)$ is directly proportional to the modal displacements ϕ_n. This indicates that stories experiencing maximum modal displacements will have

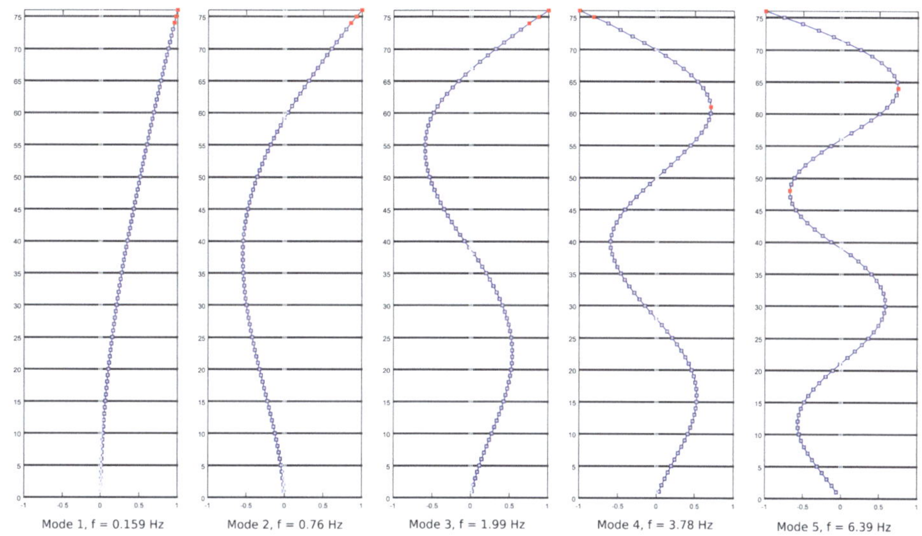

Fig. 5.21 Three stories with maximum modal displacements in first five natural modes

a heightened influence on the building's overall modal response. Moreover, these modal responses in the nth mode are intrinsically linked to the frequency content of the seismic excitation, $\ddot{u}_g$. Specifically, a greater acceleration amplitude at a given mode's frequency will yield larger modal responses.

Consequently, a particular story will exert greater influence on the total building response under the following conditions:

1. The story experiences maximum displacement in modes characterized by a high modal participation factor.
2. The story has maximum displacement in the nth mode, which may have a lower participation factor. However, this is counterbalanced by the earthquake ground motion having a larger Fourier transform amplitude at the frequency $f_n$corresponding to that mode.

These insights corroborate the validity of installing TMDs at stories other than the topmost levels, a conclusion that aligns with our empirical results.

5.2.18 Key Conclusions and Observations

Based on the empirical data collected, we can delineate several crucial conclusions:

Fig. 5.22 Fourier transform of excitations

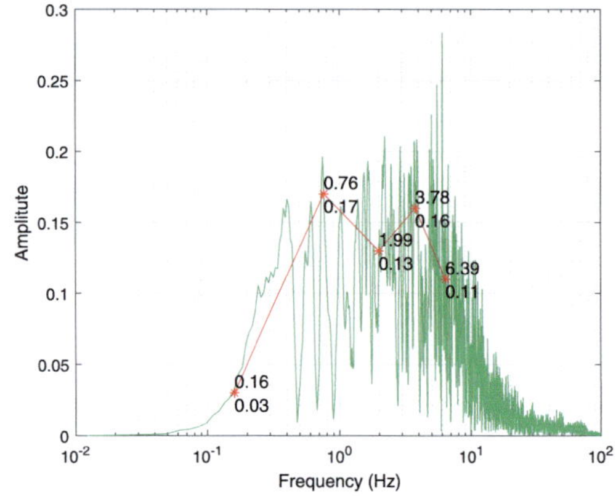

(a) Bam

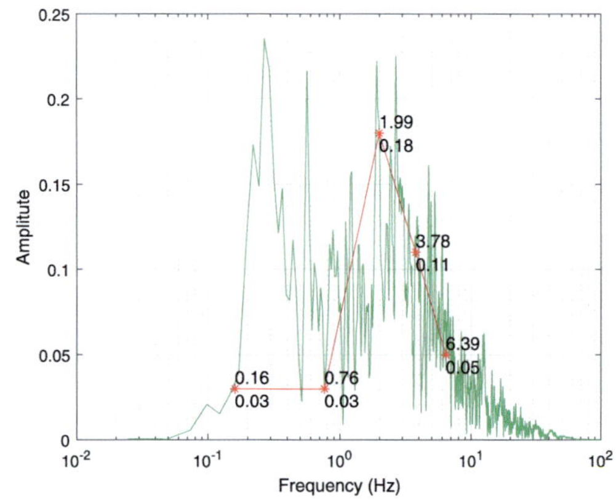

(b) Elcentro

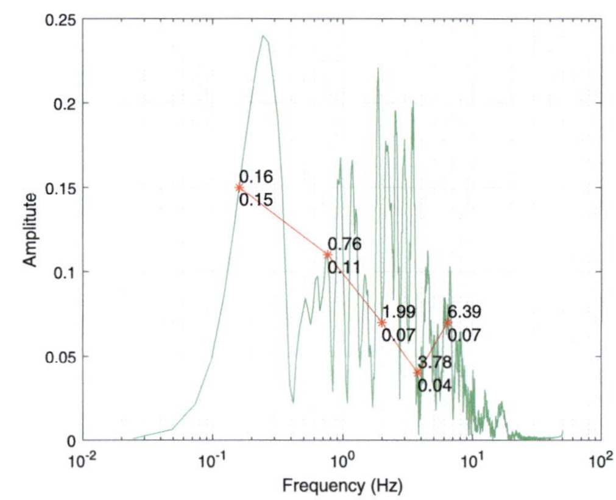

(c) Kobe

Fig. 5.22 (continued)

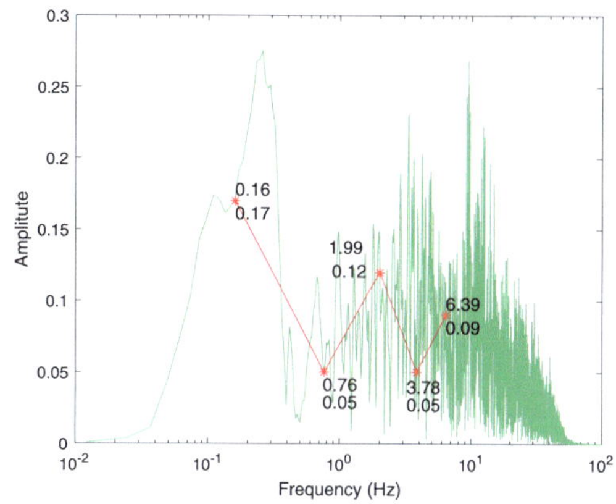

(d) Landers

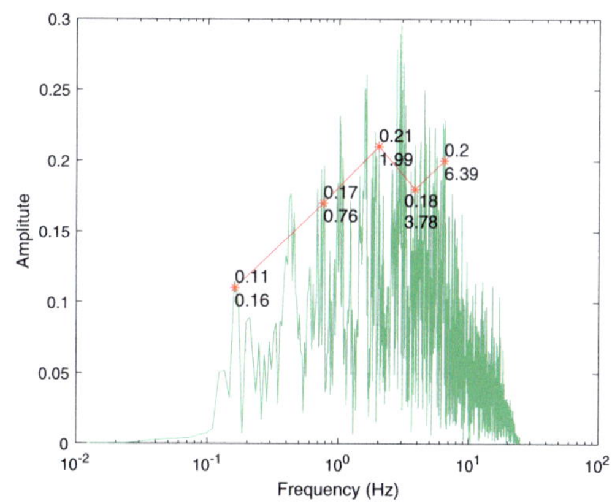

(e) Manjil

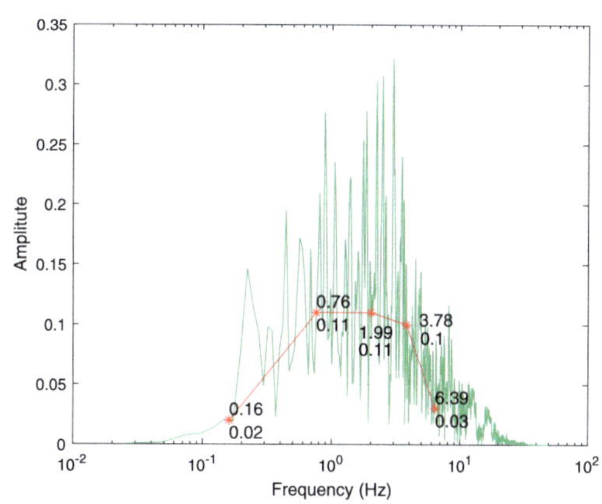

(f) Northridge

Fig. 5.22 (continued)

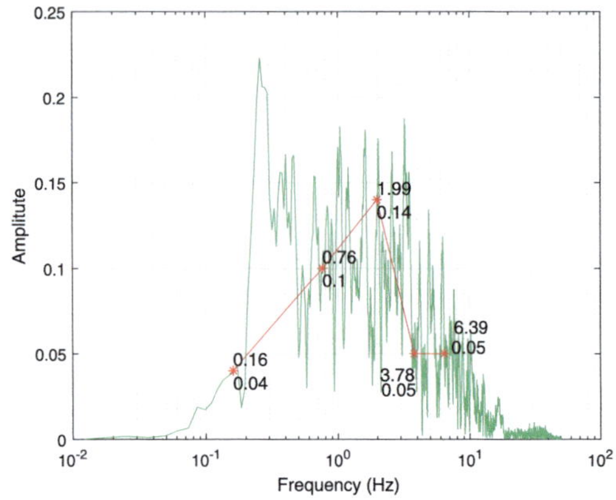

(g) San Fernando

1. Under earthquakes excitations with noticeable amplitude at structure's higher mode frequencies, Compared to a single TMD on the roof level, a distributed multiple- TMDs system is more efficient in improving structural responses with the same mass.
2. The optimum placement of the TMDs include:

   - The stories with maximum modal displacements in the lower structural modes.
   - When earthquake has a high amplitude in the higher modes of the structure, the stories corresponding to the maximum modal displacement in that modes.

3. The optimum parameters for the TMDs that control the vibrations in the lower modes include the maximum allowed damping ratio. This indicates that increasing the damping ratio would improve the performance of such TMDs.
4. The results show that the performance of the TMDs is not good in reducing the initial maximums in displacement responses compared to their reduction of the later maximums after some initial oscillations.
5. Even in the cases with immediate maximum displacement responses, the multiple distributed TMD system significantly improves the damping.

Appendix A
Algorithms and Codes

A.1 Framework Basic Algorithm

The basic algorithm of the framework includes a major loop in which the generations are evolved. In the first iteration, the algorithm generates random chromosomes to form an initial population. In each iteration, the algorithm adds new chromosomes by performing genetic operations on the current members and pass them to the *analyze* module to determine the controlled responses of the structure. After that, the *fitness* module, determines the fitness values for each chromosome and pass the population to NSGA-II module which utilizes a refined history of the TMD arrangements and corresponding fitness values, to develop the Pareto and sort the chromosomes upon that.

H. R. Rahmani, *Seismic Control of Structures Using Tuned Mass Dampers*, Synthesis Lectures on Engineering, Science, and Technology, https://doi.org/10.1007/978-3-031-71217-3

Algorithm A.1 Optimization using Genetic Algorithm

```
Initialize
population ← INITIALIZEPOPULATION()
generationNumber ← 0
Main Loop
while generationNumber ≤ N do
    generationNumber ← generationNumber + 1
    tempPopulation ← []
    Individual Processing
    while size(population) ≤ M do
        Selection
        {parent1, parent2} ← MAKESELECTION(population)
        Crossover
        if rnd(1) ≥ crossoverProbability then
            {offspring1, offspring2} ← CROSSOVER(parent1, parent2)
            Repair(offspring1, offspring2)
            COMPUTEFITNESS(offspring1, offspring2)
        end if
        Mutation
        if rnd(1) ≥ mutationProbability then
            {offspring1, offspring2} ← MUTATION(parent1, parent2)
            Repair(offspring1, offspring2)
            COMPUTEFITNESS(offspring1, offspring2)
        end if
    end while
    Pareto Front and Crowding Distance Calculation
    while size(tempPopulation) > 2 do
        p ← PARETOFRONT(tempPopulation)
        ps ← CROWDINGDISTANCE(p)
        newPopulation ← newPopulation + ps
        tempPopulation ← tempPopulation − ps
    end while
    Update Population
    population ← newPopulation
end while
```

A.2 Adapted Crossover Algorithm

The crossover function is designed to generate new offspring based on the parent chromosomes, utilizing two distinct variants:

Variant 1: Parameter Optimization: This version employs the k-point crossover method to manipulate individual TMD parameters such as stiffness, mass, and damping. During each invocation, the crossover function selectively swaps corresponding parameters between parent chromosomes. As a result, the offspring inherit TMDs situated in the same structural locations as their parents but exhibit different parametric characteristics. This approach is particularly useful for fine-tuning the TMD parameters without altering their placements.

Variant 2: Location Optimization: Unlike the first variant, this method focuses on the genetic elements corresponding to the TMD locations within the structure. It generates offspring with TMDs that retain their parents' parameters but are positioned differently within the building layout. This strategy aims to explore the optimal positioning of TMDs while keeping their specific parameters constant.

Algorithm A.2 Crossover Procedure

```
procedure CROSSOVER(individual1, individual2)
    if rnd(1) < 0.5 then
        rndTmd1 ← RANDOMSELECT TMD in individual1
        rndTmd2 ← RANDOMSELECT TMD in individual2
        k ← rndint(3)
        tmd1New ← K-POINTCROSSOVER(rndTmd1)
        tmd2New ← K-POINTCROSSOVER(rndTmd2)
        offspring1 ← REBUILD(individual1, tmd1New)
        offspring2 ← REBUILD(individual2, tmd2New)
    else
        rndTmd1 ← RANDOMSELECT TMD in individual1
        rndTmd2 ← RANDOMSELECT TMD in individual2
        tmd1New ← rndTmd2
        tmd2New ← rndTmd1
        offspring1 ← REBUILD(individual1, tmd1New)
        offspring2 ← REBUILD(individual2, tmd2New)
    end if
    return offspring1, offspring2
end procedure
```

A.3 Adapted Mutation Algorithm

The two-variant mutation function is developed to affect the (1) genes related to TMD parameters and (2) set of genes which are related to the location of the TMDs.

Algorithm A.3 Adapten Mutation Operator

```
procedure MUTATION(individual)
    if rnd(1) < 0.5 then
        rndTmd ← RANDOMSELECT TMD in individual
        newStory ← rndint(76)
        tmdNew ← MOVETMD(rndTmd, newStory)
        offspring ← REBUILD(individual, tmdNew)
    else
        rndTmd ← RANDOMSELECT TMD in individual
        targetGenCount ← rndint(10)
        for i = 1 to targetGenCount do
            targetGen ← RANDOMSELECTGEN(rndTmd)
            mutTmd ← BINARYMUTATION(rndTmd, targetGen)
        end for
        offspring ← REBUILD(individual, tmdNew)
    end if
    return offspring
end procedure
```

A.4 Matlab Code for Calculating Damping Matrix Using Rayleigh Damping

```
function [C] = Rayleigh(M, K, Modal_damp)
    % Rayleigh: Computes the Rayleigh damping matrix.
    % Input:
    %   M           - Mass matrix
    %   K           - Stiffness matrix
    %   Modal_damp  - Modal damping ratio
    % Output:
    %   C           - Damping matrix
    % Compute eigenvectors and eigenvalues of the generalized
        eigenvalue problem
    [v, d] = eig(K/M);
    % Calculate and sort natural frequencies
    Omegas = sqrt(diag(d));
    Omegas = sort(Omegas);
    % Extract the first and second natural frequencies
    w1 = Omegas(1);
    if length(Omegas) < 2
        w2 = 0; % Assign zero if only one natural frequency
            is available
    else
        w2 = Omegas(2);
    end
```

```
    % Calculate coefficients for the damping matrix
    Cmat1 = 0.5 * [1/w1, w1; 1/w2, w2];
    Modal_damps = [Modal_damp, Modal_damp]';
    alfas = inv(Cmat1) * Modal_damps;
    % Extract the derived alpha coefficients
    a0 = alfas(1, 1);
    a1 = alfas(2, 1);
    % Compute the Rayleigh damping matrix C
    C = a0 * M + a1 * K;
end
```

A.5 MATLAB CODE FOR NEWMARK-β METHOD

```
function [D, V, A] = Newmark(K, M, P, time, dt, Gamma, Beta,
    C)
    % ------------------------------------------------------------
    %| Newmark's Beta method for Linear Systems
    % ------------------------------------------------------------
    % Code by: Hamid Radmard Rahmani
    % 2015 - Germany
    % Email: Radmard.Rahmani@Gmail.com
    % ------------------------------------------------------------
    % | Reference: Dynamics of Structures 4th edition, Anil K
      . Chopra
    % ------------------------------------------------------------
    % INPUT variables:
    % K     : Stiffness matrix
    % M     : Mass matrix
    % P     : Load vector/matrix
    % time  : Total time of analysis
    % dt    : Time step
    % Gamma : Newmark's Gamma factor
    % Beta  : Newmark's Beta factor
    % C     : Damping matrix
    %
    % OUTPUT variables:
    % D     : Displacement matrix
    % V     : Velocity matrix
    % A     : Acceleration matrix
    % ------------------------------------------------------------

    % Initialize dimensions and time vector
    Dofs = size(K, 1);
    t = 0:dt:time;
    steps = length(t);

```

```
    % Initialize displacement (D), velocity (V), and
        acceleration (A) matrices
    [D, V, A] = deal(zeros(Dofs, steps));

    % Zero-padding the Load vector/matrix to match steps
    P(:, end+1:steps) = 0;

    % Constants used in Newmark's integration
    a1 = Gamma / (Beta * dt);
    a2 = 1 / (Beta * dt^2);
    a3 = 1 / (Beta * dt);
    a4 = Gamma / Beta;
    a5 = 1 / (2 * Beta);
    a6 = (Gamma / (2 * Beta) - 1) * dt;
    a7 = 1 - Gamma / Beta;

    % Modified stiffness matrix
    K_hat = M * a2 + C * a1 + K;

    % Time-stepping loop
    for i = 1:steps - 1
        % Calculate effective load for the next step
        P_hat = P(:, i+1) + (a2 * M + a1 * C) * D(:, i) + (a3
            * M + (a4 - 1) * C) * V(:, i) + ((a5 - 1) * M +
            a6 * C) * A(:, i);

        % Update displacement for the next step
        D(:, i+1) = K_hat \ P_hat;
        % Update velocity and acceleration for the next step
        V(:, i+1) = a1 * (D(:, i+1) - D(:, i)) + a7 * V(:, i)
            - dt * a6 * A(:, i);
        A(:, i+1) = a2 * (D(:, i+1) - D(:, i)) - a3 * V(:, i)
            - (a5 - 1) * A(:, i);
    end
end
```

A.6 Fast Fourier Transformation of Earthquake Record

```
% Fast Fourier Transformation (FFT) of Earthquake Record
% This script analyzes the frequency content of an earthquake record
% by applying Fast Fourier Transform (FFT) to the acceleration data.
% Import data (assuming data matrix has column 1 = time, column 2 = acceleration)
data = Elcentro;
time = data(:, 1);
acceleration = data(:, 2);

% Calculate time step, assuming uniform sampling
dt = time(2) - time(1);

% Apply FFT
n = length(acceleration);                % Length of the signal
fhat = fft(acceleration, n);             % Compute FFT
PSD = fhat .* conj(fhat) / n;            % Calculate power spectrum (power per frequency)

% Create frequency axis freq = (1 / (dt * n)) * (0:n); % X-axis of frequencies in Hz
L = 1:floor(n / 2);                      % Only plot the first half of frequencies

% Plotting the Results
%------------------------------
% Calculate a threshold (1% of max power)
threshold = 0.01 * max(PSD(L));

% Isolate frequencies above threshold
valid_frequencies = freq(L(PSD(L) > threshold));

% Find local maxima in the valid frequency range
[peaks, locs] = findpeaks(PSD(L(PSD(L) > threshold)), 'SortStr', 'descend');

% Limit frequency axis in plot
max_plot_freq = valid_frequencies(end);

% Plot the power spectrum along with significant peaks figure;
plot(freq(L(freq(L) <= max_plot_freq)), PSD(L(freq(L) <= max_plot_freq)))
hold on; plot(valid_frequencies(locs(1:3)), peaks(1:3), 'ro');

% Annotate peaks on the plot
```

```
for i = 1:min(3, length(peaks))
    text(valid_frequencies(locs(i)), peaks(i), num2str(peaks(i)), 'VerticalAlignment', 'bottom', 'HorizontalAlignment', 'right')
end
hold off;
title('Power Spectrum of Earthquake Acceleration');
xlabel('Frequency (Hz)');
ylabel('Power');
legend('Power Spectrum', 'Peaks');
grid on;

% Display and Save Results
%------------------------------------
% Create a table of peak frequencies and their corresponding power values
peakTable = table(valid_frequencies(locs(1:3))', peaks(1:3), 'VariableNames', {'Frequency_Hz', 'Power'});

% Display the table in the console
disp('Top 3 Peak Frequencies (Hz) and their Power:'); disp(peakTable);

% Save the figure to an EPS file
saveas(gcf, 'PowerSpectrum.eps', 'epsc')
```

References

1. Herrlich-media, *Tuned Mass Dampers (TMD)* (2020). https://www.lisega.de/en/vibration-control-en/tuned-mass-dampers-tmd/
2. P. Dallard, A.J. Fitzpatrick, A. Flint, S. Le Bourva, A. Low, R.M. Ridsdill Smith, M. Willford, The london millennium footbridge. Struct. Eng. **79**(22), 17–21 (2001)
3. *Tuned Mass Dampers for Railways*. https://www.wal.hk/#!damper
4. Tallest 10 completed buildings with dampers. CTBUH J. Issue III, 48 (2018)
5. A.K. Chopra, *Dynamics of Structures: Theory and Applications to Earthquake Engineering* (Always Learning, Pearson Education Limited, 2014)
6. O. Kramer, *Genetic Algorithm Essentials*, 1st edn. (Springer Publishing Company, Incorporated, 2017)
7. B.F. Spencer Jr, R.E. Christenson, S.J. Dyke, Next generation benchmark control problem for seismically excited buildings, in *Proceedings of the Second World Conference on Structural Control*, vol. 2 (Kyoto Japan, 1998), pp. 1135–1360
8. PEER Ground Motion Database–PEER Center. https://ngawest2.berkeley.edu/
9. J.T. Yao, Concept of structural control. J. Struct. Divi. **98**(st 7) (1972)
10. B.F. Spencer, M.K. Sain, Controlling buildings: A new frontier in feedback. IEEE Control Syst. Mag. **17**(6), 19–35 (1997)
11. T.T. Soong, B.F. Spencer, Supplemental energy dissipation: State-of-the-art and state-of-the-practice. Eng. Struct. **24**(3), 243–259 (2002)
12. F. Akira, S. Tomoo, S. Keiji, Control effect for 20-story benchmark building using passive or semiactive device. J. Eng. Mech. **130**(4), 430–436 (2004)
13. C. Li, Multiple active-passive tuned mass dampers for structures under the ground acceleration. Earthquake Eng. Struct. Dyn. **32**(6), 949–964 (2003)
14. M.K. Shrimali, S.D. Bharti, S.M. Dumne, Seismic response analysis of coupled building involving MR damper and elastomeric base isolation. Ain Shams Eng. J. **6**(2), 457–470 (2015)
15. O. Abdeljaber, O. Avci, D.J. Inman, Active vibration control of flexible cantilever plates using piezoelectric materials and artificial neural networks. J. Sound Vibr. **363**, 33–53 (2016)
16. H.J. Li, S.-L. James Hu, C. Jakubiak, H2 active vibration control for offshore platform subjected to wave loading. J. Sound Vibr. **263**(4), 709–724 (2003)
17. K. Bani-Hani, J. Ghaboussi, Neural networks for structural control of a benchmark problem, active tendon system. Earthquake Eng. Struct. Dyn. **27**(11), 1225–1245 (1998)
18. Gawronski, *Advanced Dynamics and Active Control of Structures* (Springer)

H. R. Rahmani, *Seismic Control of Structures Using Tuned Mass Dampers*, Synthesis Lectures on Engineering, Science, and Technology, https://doi.org/10.1007/978-3-031-71217-3

19. J. Ghaboussi, A. Joghataie, *Active control of structures using neural networks* (J. Eng, Mech, 1995)
20. M.P. Singh, E.E. Matheu, L.E. Suarez, Active and semi-active control of structures under seismic excitation. Earthquake Eng. Struct. Dyn. **26**(2), 193–213
21. L.-L. Chung, Y.-A. Lai, C.-S. Walter Yang, K.-H. Lien, L.-Y. Wu, Semi-active tuned mass dampers with phase control. J. Sound Vibr. **332**(15), 3610–3625 (2013)
22. R.P. Eason, C. Sun, A.J. Dick, S. Nagarajaiah, Attenuation of a linear oscillator using a nonlinear and a semi-active tuned mass damper in series. J. Sound Vibr. **332**(1), 154–166 (2013)
23. C.-C. Lin, L. Lyan-Ywan, G.-L. Lin, T.-W. Yang, Vibration control of seismic structures using semi-active friction multiple tuned mass dampers. Eng. Struct. **32**(10), 3404–3417 (2010)
24. N.R. Fisco, H. Adeli, Smart structures: Part II – Hybrid control systems and control strategies. Sci. Iranica **18**(3), 285–295 (2011)
25. H. Frahm, Device for damping vibrations of bodies (1911)
26. Q. Li, J. Fan, J. Nie, Q. Li, Y. Chen, Crowd-induced random vibration of footbridge and vibration control using multiple tuned mass dampers. J. Sound Vibr. **329**(19), 4068–4092 (2010)
27. S.J. Moon, L.A. Bergman, P.G. Voulgaris, Sliding mode control of cable-stayed bridge subjected to seismic excitation. J. Eng. Mech. **129**(1), 71–78 (2003)
28. N. Hoang, Y. Fujino, P. Warnitchai, Optimal tuned mass damper for seismic applications and practical design formulas. Eng. Struct. **30**(3), 707–715 (2008)
29. A. Casalotti, A. Arena, W. Lacarbonara, Mitigation of post-flutter oscillations in suspension bridges by hysteretic tuned mass dampers. Eng. Struct. **69**, 62–71 (2014)
30. J.F. Wang, C.C. Lin, B.L. Chen, Vibration suppression for high-speed railway bridges using tuned mass dampers. Int. J. Solids Struct. **40**(2), 465–491 (2003)
31. J.L. Almazán, G. Espinoza, J.J. Aguirre, Torsional balance of asymmetric structures by means of tuned mass dampers. Eng. Struct. **42**, 308–328 (2012)
32. G. Chen, W. Jingning, Experimental study on multiple tuned mass dampers to reduce seismic responses of a three-storey building structure. Earthquake Eng. Struct. Dyn. **32**(5), 793–810 (2003)
33. Y. Arfiadi, M.N.S. Hadi, Optimum placement and properties of tuned mass dampers using hybrid genetic algorithms. Iran Univ. Sci. Technol. **1**(1), 167–187 (2011)
34. S. Elias, V. Matsagar, Distributed multiple tuned mass dampers for wind vibration response control of high-rise building. J. Eng. **1–11**, 2014 (2014)
35. Y. Fujino, M. Abe, Design formulas for tuned mass dampers based on a perturbation technique. Earthquake Eng. Struct. Dyn. **22**(10), 833–854 (1993)
36. K.C.S. Kwok, B. Samali, Performance of tuned mass dampers under wind loads. Eng. Struct. **17**(9), 655–667 (1995)
37. A.Y.T. Leung, H. Zhang, Particle swarm optimization of tuned mass dampers. Eng. Struct. **31**(3), 715–728 (2009)
38. C. Li, Performance of multiple tuned mass dampers for attenuating undesirable oscillations of structures under the ground acceleration. Earthquake Eng. Struct. Dyn. **29**(9), 1405–1421 (2000)
39. C.-L. Lee, Y.-T. Chen, L.-L. Chung, Y.-P. Wang, Optimal design theories and applications of tuned mass dampers. Eng. Struct. **28**(1), 43–53 (2006)
40. Y.L. Xu, J. Chen, C.L. Ng, W.L. Qu, Semiactive seismic response control of buildings with podium structure. J. Struct. Eng. **131**(6), 890–899 (2005)
41. O. Araz, V. Kahya, Optimization of multiple tuned mass dampers for a two-span continuous railway bridge via differential evolution algorithm, in *Structures*, vol. 39 (Elsevier, 2022), pp. 29–38
42. A. Bozer, ŞS. Özsarıyıldız, Free parameter search of multiple tuned mass dampers by using artificial bee colony algorithm. Struct. Control Health Monitor. **25**(2), e2066 (2018)

43. N. Carpineto, W. Lacarbonara, F. Vestroni, Mitigation of pedestrian-induced vibrations in suspension footbridges via multiple tuned mass dampers. J. Vibr. Control **16**(5), 749–776 (2010)
44. Y. Guo, W. Chen, Dynamic analysis of space structures with multiple tuned mass dampers. Eng. Struct. **29**(12), 3390–3403 (2007)
45. N. Hoang, P. Warnitchai, Design of multiple tuned mass dampers by using a numerical optimizer. Earthquake Eng. Struct. Dyn. **34**(2), 125–144 (2005)
46. G. Chen, W. Jingning, Optimal placement of multiple tune mass dampers for seismic structures. J. Struct. Eng. **127**(9), 1054–1062 (2001)
47. T. Igusa, X. Kangming, Wide-band response of multiple subsystems with high modal density, in *Stochastic Structural Dynamics 1*. ed. by Y.K. Lin, I. Elishakoff (Springer, Berlin Heidelberg, 1991), pp.131–145
48. M. Jokic, M. Stegic, M. Butkovic, Reduced-order multiple tuned mass damper optimization: A bounded real lemma for descriptor systems approach. J. Sound Vibr. **330**(22), 5259–5268 (2011)
49. A.S. Joshi, R.S. Jangid, Optimum parameters of multiple tuned mass dampers for base-excited damped systems. J. Sound Vibr. **202**(5), 657–667 (1997)
50. A. Kareem, Performance of multiple mass dampers under random loading, p. 14
51. K. KrishnaKumar, R. Swaminathan, L. Montgomery, Multiple optimal solutions for structural control using genetic algorithms with niching. J. Guid. Control Dyn. **17**(6), 1374–1377 (1994)
52. C. Li, Y. Liu, Active multiple tuned mass dampers for structures under the ground acceleration. Earthquake Eng. Struct. Dyn. **31**(5), 1041–1052 (2002)
53. C. Li, Q. Weilian, Evaluation of elastically linked dashpot based active multiple tuned mass dampers for structures under ground acceleration. Eng. Struct. **26**(14), 2149–2160 (2004)
54. C.C. Lin, G.L. Lin, H.Y. Lung, Dynamic test of multiple tuned mass dampers for vibration control of high-rise buildings, p. 11
55. C. Li, B. Zhu, Estimating double tuned mass dampers for structures under ground acceleration using a novel optimum criterion. J. Sound Vibr. **298**(1–2), 280–297 (2006)
56. T.A. Sakr, Vibration control of buildings by using partial floor loads as multiple tuned mass dampers. HBRC J
57. Assessment of vibrational behaviour of footbridges under pedestrian loading, technical guide. *Technical Department for Transport, Roads and Bridges Engineering and Road Safety (SETRA)* (Ministry of Transport, Republique Francaise, 2006)
58. J. Brownjohn, P. Fok, M. Roche, P. Omenzetter, Long span steel pedestrian bridge at singapore changi airport. part 2: Crowd loading tests and vibration mitigation measures (2004)
59. E. Caetano, Á. Cunha, C. Moutinho, F. Magalhães, Studies for controlling human-induced vibration of the pedro e inês footbridge, portugal. part 2: Implementation of tuned mass dampers. Eng. Struct. **32**(4), 1082–1091 (2010)
60. K. Van Nimmen, P. Verbeke, G. Lombaert, G. De Roeck, P. Van den Broeck, Numerical and experimental evaluation of the dynamic performance of a footbridge with tuned mass dampers. J. Bridge Eng. **21**(8), C4016001 (2016)
61. D. Wang, W. Chengqing, Y. Zhang, S. Li, Study on vertical vibration control of long-span steel footbridge with tuned mass dampers under pedestrian excitation. J. Construct. Steel Res. **154**, 84–98 (2019)
62. M. Behloul, K.C. Lee, Ductal® seonyu footbridge. Struct. Concr. **4**(4), 195–201 (2003)
63. J.-D. Yau, Train-induced vibration control of simple beams using string-type tuned mass dampers. J. Mech. **23**(4), 329–340 (2007)
64. J. Li, S. Mubiao, L. Fan, Vibration control of railway bridges under high-speed trains using multiple tuned mass dampers. J. Bridge Eng. **10**(3), 312–320 (2005)

65. V. Kahya, O. Araz, A simple design method for multiple tuned mass dampers in reduction of excessive vibrations of high-speed railway bridges. J. Facul. Eng. Architect. Gazi Univ. **35**(2), 607–618 (2020)
66. O. Araz, V. Kahya, Series tuned mass dampers in vibration control of continuous railway bridges. Struct. Eng. Mech. An Int'l J. **73**(2), 133–141 (2020)
67. C.C. Lin, J.-F. Wang, B.L. Chen, Train-induced vibration control of high-speed railway bridges equipped with multiple tuned mass dampers. J. Bridge Eng. **10**(4), 398–414 (2005)
68. M. Luu, V. Zabel, C. Könke, An optimization method of multi-resonant response of high-speed train bridges using tmds. Fin. Elements Anal. Design **53**, 13–23 (2012)
69. A. Andersson, A. O'Connor, R. Karoumi, Passive and adaptive damping systems for vibration mitigation and increased fatigue service life of a tied arch railway bridge. Comput. Aided Civil Infrastruct. Eng. **30**(9), 748–757 (2015)
70. X. Chen, A. Kareem, Efficacy of tuned mass dampers for bridge flutter control. J. Struct. Eng. **129**(10), 1291–1300 (2003)
71. S.-D. Kwon, K.-S. Park, Suppression of bridge flutter using tuned mass dampers based on robust performance design. J. Wind Eng. Ind. Aerodyn. **92**(11), 919–934 (2004)
72. A. Casalotti, A. Arena, W. Lacarbonara, Mitigation of post-flutter oscillations in suspension bridges by hysteretic tuned mass dampers. Eng. Struct. **69**, 62–71 (2014)
73. F. Ubertini, G. Comanducci, S. Laflamme, A parametric study on reliability-based tuned-mass damper design against bridge flutter. J. Vibr. Control **23**(9), 1518–1534 (2017)
74. Y.-Y. Lin, C.-M. Cheng, D. Sun et al., Wind-induced vibration control of long-span bridges by multiple tuned mass dampers. J. Appl. Sci. Eng. **3**(1), 1–13 (2000)
75. Z. Sun, Z. Zou, X. Ying, X. Li, Tuned mass dampers for wind-induced vibration control of chongqi bridge. J. Bridge Eng. **25**(1), 05019014 (2020)
76. X. Yin, G. Song, Y. Liu, Vibration suppression of wind/traffic/bridge coupled system using multiple pounding tuned mass dampers (mptmd). Sensors **19**(5), 1133 (2019)
77. N. Debnath, S.K. Deb, A. Dutta, Multi-modal vibration control of truss bridges with tuned mass dampers under general loading. J. Vibr. Control **22**(20), 4121–4140 (2016)
78. H. Niu, Z. Chen, X. Hua, W. Zhang, Mitigation of wind-induced vibrations of bridge hangers using tuned mass dampers with eddy current damping. Smart Struct. Syst. Int. J. **22**(6), 727–741 (2018)
79. J.R. Sladek, R.E. Klingner, Effect of tuned-mass dampers on seismic response. J. Struct. Eng. **109**(8), 2004–2009 (1983)
80. R.J. McNamara, Tuned mass dampers for buildings. J. Struct. Divis. **103**(9), 1785–1798 (1977)
81. F. Sadek, B. Mohraz, A.W. Taylor, R.M. Chung, A method of estimating the parameters of tuned mass dampers for seismic applications. Earthquake Eng. Struct. Dyn. **26**(6), 617–635 (1997)
82. R.W. Luft, Optimal tuned mass dampers for buildings. J. Struct. Divis. **105**(12), 2766–2772 (1979)
83. A.M. Kaynia, D. Veneziano, J.M. Biggs, Seismic effectiveness of tuned mass dampers. J. Struct. Divis. **107**(8), 1465–1484 (1981)
84. M.-Y. Liu, W.-L. Chiang, J.-H. Hwang, C.-R. Chu, Wind-induced vibration of high-rise building with tuned mass damper including soil-structure interaction. J. Wind Eng. Ind. Aerodyn. **96**(6–7), 1092–1102 (2008)
85. A. Mousaad Aly, Proposed robust tuned mass damper for response mitigation in buildings exposed to multidirectional wind. Struct. Design Tall Special Build. **23**(9), 664–691 (2014)
86. L. Zuo, S.A. Nayfeh, Optimization of the individual stiffness and damping parameters in multiple-tuned-mass-damper systems. J. Vibr. Acoust. **127**(1), 77 (2005)
87. A.Y. Pisal, R.S. Jangid, Seismic response of multi-story structure with multiple tuned mass friction dampers. Int. J. Adv. Struct. Eng. (IJASE) **6**(1), 1–13 (2014)

88. C. Li, Q. Weilian, Optimum properties of multiple tuned mass dampers for reduction of translational and torsional response of structures subject to ground acceleration. Eng. Struct. **28**(4), 472–494 (2006)
89. C.C. Chang, W.L. Qu, Unified dynamic absorber design formulas for wind-induced vibration control of tall buildings. Struct. Design Tall Build. **7**(2), 147–166 (1998)
90. S.N. Deshmukh, N.K. Chandiramani, LQR control of wind excited benchmark building using variable stiffness tuned mass damper. Shock Vibr. **1–12**, 2014 (2014)
91. S. Pourzeynali, S. Salimi, H. Eimani Kalesar, Robust multi-objective optimization design of TMD control device to reduce tall building responses against earthquake excitations using genetic algorithms. Sci. Iranica **20**(2), 207–221 (2013)
92. S. Elias, V. Matsagar, T.K. Datta, Effectiveness of distributed tuned mass dampers for multi-mode control of chimney under earthquakes. Eng. Struct. **124**, 1–16 (2016)
93. A comparative analysis of selection schemes used in genetic algorithms. Found. Gen. Algorith. **1**, 69–93 (1991)
94. K. Deb, A. Pratap, S. Agarwal, T. Meyarivan, A fast and elitist multiobjective genetic algorithm: NSGA-II. IEEE Trans. Evol. Comput. **6**(2), 182–197 (2002)
95. J.N. Yang, B. Samali, J. Wu, Benchmark problem for response control wind excited tall buildings. J. Eng. Mech. **130**(4), 437–446 (2004)
96. H. Radmard Rahmani, C. Könke, Seismic control of tall buildings using distributed multiple tuned mass dampers. Adv. Civil Eng. **1–19**, 2019 (2019)
97. S. Elias, V. Matsagar, T.K. Datta, Distributed tuned mass dampers for multi-mode control of benchmark building under seismic excitations. J. Earthquake Eng. 1–36 (2017)
98. H.-N. Li, X.-L. Ni, Optimization of non-uniformly distributed multiple tuned mass damper. J. Sound Vibr. **308**(1), 80–97 (2007)
99. G. Carlo Marano, R. Greco, B. Chiaia, A comparison between different optimization criteria for tuned mass dampers design. J. Sound Vibr. **329**(23), 4880–4890 (2010)
100. S. Elias, V. Matsagar, Research developments in vibration control of structures using passive tuned mass dampers. Ann. Rev. Control **44**, 129–156 (2017)
101. G. Barratt Warburton, Optimum absorber parameters for various combinations of response and excitation parameters. Earthquake Eng. Struct. Dyn. **10**(3), 381–401 (1982)
102. T. Ioi, K. Ikeda, On the dynamic vibration damped absorber of the vibration system. Bull. Japanese Soc. Mech. Eng. **21**(151), 64–71 (1978)
103. G.B. Warburton, Optimum absorber parameters for various combinations of response and excitation parameters. Earthquake Eng. Struct. Dyn. **10**(3), 381–401 (1982)
104. J.P. Den Hartog, *Mechanical Vibrations* (McGraw-Hill, N.Y., 1956)
105. R.G. Jacquot, D.L. Hoppe, Optimal random vibration absorbers. J. Eng. Mech. ASCE **99**, 612–616 (1973)
106. G.B. Warburton, Optimal absorber parameters for various combinations of response and excitation parameters. Earthquake Eng. Struct. Dyn. **10**, 381–401 (1982)
107. K.C. Falcon, B.J. Stone, W.D. Simcock, C. Andrew, Optimization of vibration absorbers: A graphical method for use on idealized systems with restricted damping. J. Mech. Eng. Sci. **9**, 374–381 (1967)
108. R.W. Luft, Optimal tuned mass dampers for buildings. J. Struct. Divis. ASCE **105**(12), 2766–2772 (1979)
109. N.M. Newmark, A method of computation for structural dynamics. J. Eng. Mech. Divis. **85**(3), 67–94 (1959)
110. American Society of Civil Engineers, *Minimum Design Loads for Buildings and Other Structures: Third Printing* (American Society of Civil Engineers, 2013)
111. M. Gen, R. Cheng, *Genetic Algorithms and Engineering Optimization* (John Wiley & Sons, 2000)

MIX
Papier aus verantwortungsvollen Quellen
Paper from responsible sources
FSC® C105338

If you have any concerns about our products,
you can contact us on
ProductSafety@springernature.com

In case Publisher is established outside the EU,
the EU authorized representative is:
Springer Nature Customer Service Center GmbH
Europaplatz 3, 69115 Heidelberg, Germany

Printed by Libri Plureos GmbH
in Hamburg, Germany